Visually Memorable Neuroanatomy for Beginners

Visually Memorable Neuroanatomy for Beginners

Min Suk Chung

Beom Sun Chung

ELSEVIER

ACADEMIC PRESS
An imprint of Elsevier

British Library Cataloguing-in-Publication Data
A catalogue record for this book is available from the British Library

Library of Congress Cataloging-in-Publication Data
A catalog record for this book is available from the Library of Congress

ISBN: 978-0-12-819901-5

For Information on all Academic Press publications
visit our website at https://www.elsevier.com/books-and-journals

Publisher: Nikki Levy
Acquisitions Editor: Natalie Farra
Editorial Project Manager: Samantha Allard
Production Project Manager: Punithavathy Govindaradjane
Cover Designer: Miles Hitchen

Typeset by MPS Limited, Chennai, India

Working together
to grow libraries in
developing countries

www.elsevier.com • www.bookaid.org

Contents

Preface

Nowadays, neuroanatomy is learned by the countless students in medical and bioscience fields. This is because neuroanatomy is the basis of neurology, neurosurgery, neuroimaging, neurophysiology, neuropharmacology, and so on. Without knowledge of neuroanatomy, one's understanding of neuroscience would be a house of cards.

From the authors' viewpoint as anatomists, there is a question. What is the difference between neuroanatomy and regional anatomy? One answer is about the scope. Regional anatomy mainly deals with the peripheral nervous system (cranial nerve, spinal nerve), while neuroanatomy deals with both the central nervous system (brain, spinal cord) and peripheral nervous system. The central nervous system cannot be understood without the peripheral nervous system. Therefore, this neuroanatomy book delivers essential contents of the peripheral nervous system. Unsatisfied readers are suggested to learn regional anatomy. One of the choices is another book by the authors, *Visually Memorable Regional Anatomy*, which can be obtained on the website (anatomy.co.kr).

Another answer is about the neuronal connection. The nervous system can be explained in two aspects. First, the gross morphology, identifiable by cadaver dissection, is essential for comprehension of the nerve's actual appearance. Second, the neuronal connection, identifiable by special microscopic observation, is essential for comprehension of the nerve's function. Whereas the former is mainly learned in regional anatomy, the latter is intensively learned in neuroanatomy. This book contains plenty of illustrations regarding both aspects.

Students generally learn neuroanatomy with conventional textbooks. Regretfully, most students perceive neuroanatomy as a terrifying subject because of its overwhelming amount and extreme difficulty of content. In the natural course, they suffer from neurophobia.

The purpose of this book is to help students overcome their neurophobia and study neuroanatomy comfortably. Fittingly, the last two words of this book title are *for Beginners*. To serve this purpose, this book concentrates on easy-to-read stories rather than exhaustive details. The easy-to-read stories should be like the solving process for a math equation which is logically successive.

This book is neither complete nor thoroughly exact. Notes to keep in mind regarding the imperfections are as follows.

Some detailed information is not introduced, so as to make the book simpler and easier for novices. For example, we do not describe the fact that the spinal nucleus of trigeminal nerve is related to the facial nerve. Sometimes, the details are introduced with the word "exactly" in italics even though we confess the details are insufficient. The sentences in italics are not illustrated in this book.

We hardly explain numbers like those for the Brodmann area (e.g., 3, 1, 2 = postcentral gyrus) because the cerebral cortex can be comprehended without these numbers. Everyone knows that numbers are easily forgettable. Eponyms that are difficult to memorize are omitted as well. For instance, the term "medial limbic circuit" is used instead of "Papez circuit."

Clinical neuroanatomy is barely dealt with in this book. Diseases such as Parkinson disease and Huntington disease are not discussed. We have concentrated on neuroanatomy itself and its supportive embryology, rather than on clinical knowledge. It would be beneficial for the medical students to familiarize themselves with the diseases later.

The illustrations in this book are extremely simple. An example is the cover picture, where the cerebrum, thalamus, and brainstem are depicted with three simple swellings. This drawing is effective in explaining the course of sensory and motor nerves consistently. Readers can easily imitate the schematics, which is helpful for memorization. Notice the first two words of this book title, *Visually Memorable*. Comparing the schematic figures with the realistic atlas is mandatory, to gain an accurate insight.

In consideration of the small amount of neuroanatomy information even without full references, some may criticize this book as second class. However, the authors have a different idea. After grasping fundamental knowledge with this book, students can comfortably and confidently study advanced topics in neuroanatomy and other classes.

This book contains memorizing tips (mnemonics). As an example, the Lateral geniculate nucleus is for Light; the Medial geniculate nucleus is for Music. A quarter of the tips have been created by others to whom the authors are grateful. As another example, the medulla oblongata is regarded as the spinal cord (medulla) that is elongated (oblongata). Such etymology facilitates both short-term and long-term memories. Moreover, cartoons in two styles drawn by the first author are included in order to make neuroanatomy approachable. The readers may choose their favorites among the provided mnemonics and cartoons as if they were shopping.

At the end of the book, there are organized tables of the afferent nerves having three neurons. The tables, devised by the authors, show the general rule that applies to the numerous afferent nerves.

Here we faithfully follow the official terms of *Terminologia Anatomica*. However, some official terms are slightly modified (e.g., oculomotor nucleus substituting for nucleus of oculomotor nerve) and some terms are coined (e.g., dorsal sensory plate, ventral motor plate) for convenience. Also, abbreviations are utilized, which are introduced in the prologue (e.g., CN III for oculomotor nerve).

In traditional neuroanatomy books, the horizontal plane is viewed from the superior side. However, we have chosen the inferior view to offer coherence with CT and MRI. By convention, the sensory nerve is drawn in blue and the motor nerve in red.

Beom Jo Chung helped to create drawings of the book on Adobe Illustrator. Students in Ajou University School of Medicine (especially Byung Moo Kim, Jeongwon Kim, Soyeon Park) provided the suitable source of drawings and writings. Korean anatomists (In Hyuk Chung, Kyung-Seok Hu, Yonghyun Jun, Dong Woon Kim, Soonwook Kwon, Jae-Ho Lee, Won Taek Lee, Young-Don Lee, Chang-Seok Oh, Jin Seo Park, Kyung Ah Park, Gu Seob Roh, Haeyoung Suh-Kim) and Korean clinicians (Je-Geun Chi, Byung Gon Kim, Sun Ah Park, Tae Hoon Roh) provided useful suggestions and corrections. A friendly clinician (Eun Seo Kim) gave a helping hand. The main work of this book was financially supported by the Ministry of Trade, Industry and Energy (MOTIE) and Korea Institute for Advancement of Technology (KIAT) through the International Cooperative R&D program (Grant number: N0002249). The project "NEUROMAN" was carried out with Niels Kuster in IT'IS, Switzerland.

It must have been a difficult job for Natalie Farra (Acquisitions Editor) to authorize this book which has an unusual and whimsical style. Samantha Allard (Editorial Project Manager) helpfully guided the authors through all processes of book editing. The authors express their gratitude to Punithavathy Govindaradjane (Production Project Manager) and her staff for the repeated revision and excellent production of this book.

The authors wish this book to serve as a truly helpful resource to students studying neuroanatomy. Neuroanatomy should be understood concretely, not memorized blindly. Enjoying neuroanatomy is better than suffering from it.

April, 2021

Min Suk Chung, MD, PhD

Born: Seoul, South Korea (1961)

MD: Yonsei University, Seoul, South Korea (1980−87)

MS/PhD: Graduate School, Yonsei University, Seoul, South Korea (1987−96)

Visiting Scholar: Stanford University School of Medicine, California, United States (2004)

Full-Time Instructor, Assistant Professor, Associate Professor, and Professor: Department of Anatomy, Ajou University School of Medicine, Suwon, South Korea (1993−Present)

Beom Sun Chung, MD, PhD

Born: Seoul, South Korea (1989)

MD: Soonchunhyang University, Cheonan, South Korea (2008−14)

MS/PhD: Graduate School, Ajou University, Suwon, South Korea (2014−20)

Postdoctoral Fellow: Tulane Center for Clinical Neurosciences, Tulane University School of Medicine, Louisiana, United States (2020)

Assistant Professor: Department of Anatomy, Yonsei University Wonju College of Medicine, Wonju, South Korea (2021−Present)

Prologue

1. Names of the cranial nerves (CN) and spinal nerves are written in following abbreviations.

   CN I = Olfactory nerve
   CN II = Optic nerve
   CN III = Oculomotor nerve
   CN IV = Trochlear nerve
   CN V = Trigeminal nerve
      CN V1 = Ophthalmic nerve
      CN V2 = Maxillary nerve
      CN V3 = Mandibular nerve
   CN VI = Abducens nerve
   CN VII = Facial nerve
   CN VIII = Vestibulocochlear nerve
   CN IX = Glossopharyngeal nerve
   CN X = Vagus nerve
   CN XI = Accessory nerve
   CN XII = Hypoglossal nerve
   C1 = 1st cervical nerve
   T1 = 1st thoracic nerve
   L1 = 1st lumbar nerve
   S1 = 1st sacral nerve

2. For orientations of illustrations, head figures are utilized as shown below. In the case of bilateral structures, right side is depicted in most cases.

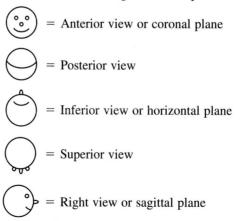

= Anterior view or coronal plane

= Posterior view

= Inferior view or horizontal plane

= Superior view

= Right view or sagittal plane

Chapter 1

Morphology of the central nervous system

The nervous system consists of the central nervous system (brain, spinal cord) and the peripheral nervous system (cranial nerve, spinal nerve). This chapter explores the gross morphology of the central nervous system, in preparation for further study of the neuronal connections. This chapter details the blood supply and cerebrospinal fluid flow of the central nervous system. Then it sequentially describes the morphology of the cerebral hemisphere, limbic system, basal nuclei, diencephalon, cerebellum, brainstem, and spinal cord. It is necessary to correlate external features of the structures to their sectional planes. It is suggested to review this chapter with other learning materials such as realistic neuroanatomy atlases, plastic specimens, three-dimensional computer models, and cadavers.

Introduction

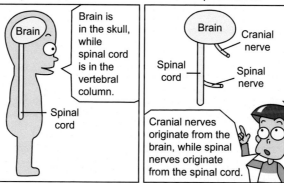

Fig. 1.1

Visually Memorable Neuroanatomy for Beginners. DOI: https://doi.org/10.1016/B978-0-12-819901-5.00001-6

The nervous system is a complex network of nerves that carry impulses between the brain, spinal cord, and various parts of the body.

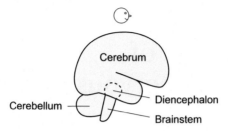

Fig. 1.2 Brain components.

When the brain is viewed laterally, its three components are identifiable: cerebrum, cerebellum, and brainstem. The diencephalon is hidden by the cerebrum (cerebral hemisphere) (Fig. 1.11).

The blood supply, the cerebrospinal fluid flow

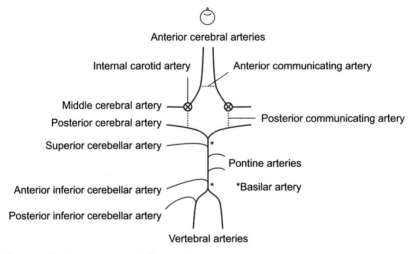

Fig. 1.3 Cerebral arteries, cerebellar arteries.

The basilar artery arises from the confluence of the two vertebral arteries at the junction between the pons and medulla oblongata. Branches of the basilar artery, named pontine arteries, feed the pons (Fig. 1.51).

The posterior inferior cerebellar artery branches off from the vertebral artery, while the anterior inferior cerebellar artery and superior cerebellar artery branch off from the basilar artery. This is because the basilar artery is on the pons (Fig. 1.54) which is right in front of the cerebellum (Figs. 1.44, 5.6).

There are three cerebral arteries as well as three cerebellar arteries on each side. The posterior cerebral artery is a terminal division of the basilar artery, while the middle and anterior cerebral arteries are two divisions of the internal carotid artery.

The "posterior" cerebral arteries and internal carotid arteries are connected by the "posterior" communicating arteries, while the bilateral "anterior" cerebral arteries are connected by the "anterior" communicating artery.

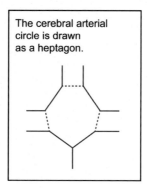

The cerebral arterial circle is drawn as a heptagon.

Fig. 1.4

The cerebral arterial circle (circle of Willis) is composed of the posterior cerebral arteries, posterior communicating arteries, anterior cerebral arteries, and anterior communicating artery *(Exactly, a short segment of internal carotid artery is included.)* (Fig. 1.3). The circle is an anastomosis that guarantees blood supply to the cerebrum.

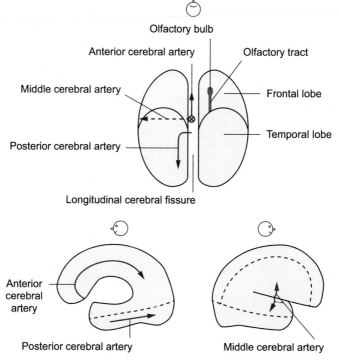

Fig. 1.5 Anterior, middle, and posterior cerebral arteries.

The anterior cerebral artery passes along the medial surface of the cerebral hemisphere anteriorly, superiorly, and then posteriorly. The middle cerebral artery emerges from the lateral sulcus to take charge of most of the lateral surface of the cerebral hemisphere (Fig. 1.23). The posterior cerebral artery passes posteriorly along its inferomedial surface (Figs. 1.6, 1.30).

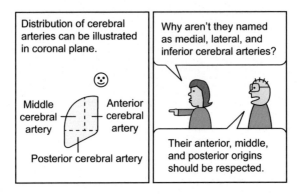

Fig. 1.6

The anterior and middle cerebral arteries supply blood to the cerebral hemisphere above a certain horizontal plane; the posterior cerebral artery feeds the cerebral hemisphere below the plane (Fig. 1.5). In other words, the horizontal plane is a territorial border between the internal carotid artery and the vertebral artery (Fig. 1.3).

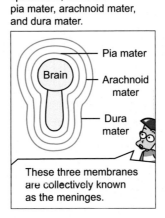

Fig. 1.7

The meninges which cover the brain and spinal cord are like PAD. The meninges are composed of Pia, Arachnoid, and Dura maters.

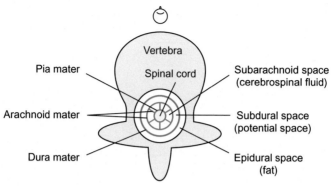

Fig. 1.8 Meninges of spinal cord.

The pia mater is adhesive to the brain (Figs. 1.14, 1.31) and spinal cord, the arachnoid (spider's) mater is entangled like a spider's web, and the dura mater is thick (Fig. 1.17).

Fig. 1.9

The DURA mater reminds us of a DURAble mother.

The subarachnoid space of brain and spinal cord is an actual space containing cerebrospinal fluid (Fig. 1.8), cerebral arteries, and cerebral veins (Fig. 1.17). In the brain, the pia mater enters the sulcus, but the arachnoid mater does not, so the subarachnoid space has substantial volume (Fig. 1.31).

Conversely, the subdural space of the brain and spinal cord is a potential space. Its volume is close to zero unlike Fig. 1.8 and Fig. 1.17, and increases in case of hemorrhage.

Whereas the epidural space of the brain is negligible (Fig. 1.17), the epidural space of the spinal cord is filled with fat (Fig. 1.8).

6

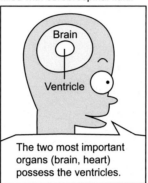

Fig. 1.10

The ventricle is cavity in the brain, where cerebrospinal fluid is produced (Fig. 1.14) and flows afterward (Figs. 1.11, 1.12).

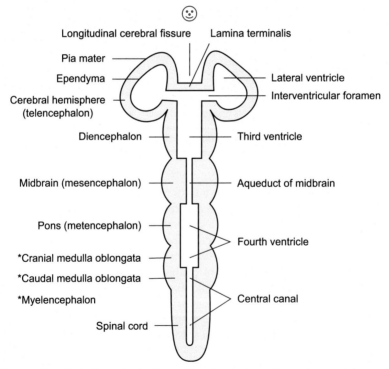

Fig. 1.11 Neural tube becoming brain, spinal cord; neural canal becoming ventricles, central canal.

The above figure shows the neural tube, which becomes the brain and spinal cord (Fig. 5.6) during embryological development. Inside of the neural tube is the neural canal (Fig. 5.7), which becomes the ventricles and central canal.

Among the serial ventricles, the largest ones are the two lateral ventricles in the right and left cerebral hemispheres. The third ventricle is between the right and left diencephalons (thalami, hypothalami) (Figs. 1.40, 1.41). The aqueduct of midbrain is literally in the midbrain (Figs. 1.44, 1.52). Instead of the aqueduct of midbrain, an incorrect term "cerebral aqueduct" is frequently used. How ridiculous!

The fourth ventricle is located in the pons (Fig. 1.54) and cranial medulla oblongata (Fig. 1.58). The central canal is situated in the caudal medulla oblongata (Fig. 1.59) and spinal cord (Figs. 1.44, 1.68). This implies that morphologically, the cranial medulla oblongata is similar to the pons; the caudal medulla oblongata is similar to the spinal cord.

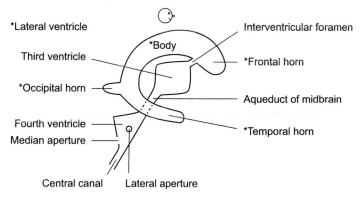

Fig. 1.12   Ventricles, central canal.

In the above figure, ventricles excluding the left lateral ventricle are presented. The frontal horn, body, and temporal horn of lateral ventricle are C-shaped.

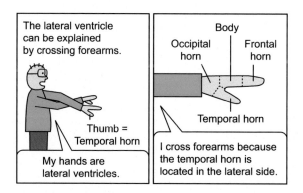

Fig. 1.13

The frontal horn extends forward into the frontal lobe; the occipital horn extends backward into the occipital lobe; the temporal horn extends forward, laterally into the temporal lobe (Figs. 1.12, 1.23, 1.30, 1.40).

Each lateral ventricle opens to the third ventricle via the interventricular foramen (Figs. 1.11, 1.40, 1.44) which is between the frontal horn and body (Fig. 1.12).

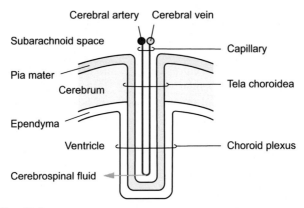

Fig. 1.14   Choroid plexus.

The cerebrum is covered by the pia mater (Fig. 1.7), while the ventricle is lined by the ependyma (Fig. 5.8). The cerebral artery in the subarachnoid space (Fig. 1.17) gives off capillary that invaginates into the cerebrum and ventricle. The invaginated capillary surrounded by the pia mater in the cerebrum is the tela choroidea; the further invaginated capillary surrounded by the pia mater and ependyma in the ventricle is the choroid plexus. At the choroid plexus, plasma in the capillary flows out to the ventricle and becomes cerebrospinal fluid.

The choroid plexus exists in the lateral, third, and fourth ventricles. The cerebrospinal fluid produced in the lateral ventricle flows through the third ventricle, aqueduct of midbrain, and fourth ventricle sequentially (Figs. 1.11, 1.12). The ordinal numbers have the meaning.

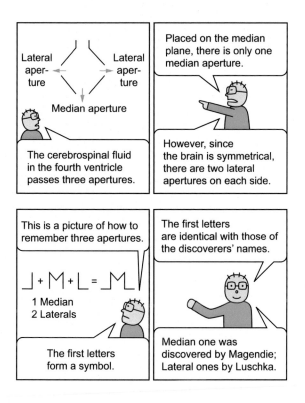

Fig. 1.15

The cerebrospinal fluid in the fourth ventricle exits to the subarachnoid space (Figs. 1.8, 1.17). The three exits are one median aperture (Figs. 1.12, 1.44) and two lateral apertures (Fig. 1.51).

Fig. 1.16

The cerebrospinal fluid surrounds and protects the brain and spinal cord which are soft and fragile (Figs. 1.8, 1.17). Additionally, the cerebrospinal fluid has diverse functions such as substance distribution and waste clearing.

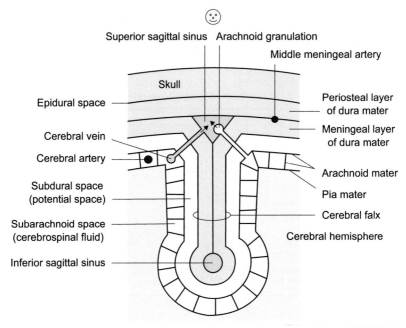

Fig. 1.17   Meninges of brain.

The dura mater of the brain consists of periosteal and meningeal layers. The above figure (coronal plane) demonstrates two components of dural venous sinuses: the superior and inferior sagittal sinuses (Fig. 1.21). The superior sagittal sinus is surrounded by the periosteal and meningeal layers, whereas the inferior one is surrounded only by the meningeal layer. Consequently, the superior sagittal sinus is in contact with the skull, whereas the inferior one is not. The cerebral falx connects the two sinuses and occupies the longitudinal cerebral fissure (Fig. 1.5).

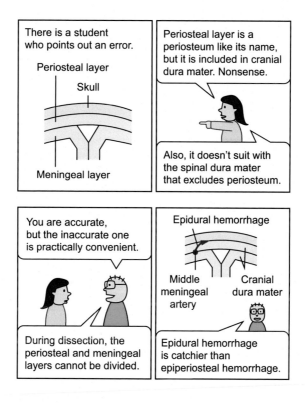

Fig. 1.18

Between the periosteal and meningeal layers, there is no recognizable space (exceptions: dural venous sinuses, middle meningeal artery). If the middle meningeal artery is ruptured by skull fracture, blood accumulates in the external space of the periosteal layer (epidural space) (Fig. 1.17).

If the cerebral vein emptying into the superior sagittal sinus is ruptured, blood accumulates in the subdural space; if the cerebral artery is ruptured, blood accumulates in the subarachnoid space (Fig. 1.17).

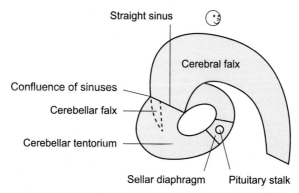

Fig. 1.19 Cerebral falx, adjacent structures.

The invaginated meningeal layers fuse to form the cerebral falx (Fig. 1.17), cerebellar falx, cerebellar tentorium, and sellar diaphragm (Fig. 1.46), which are continuous structures. Except the sellar diaphragm, they collectively meet at the straight sinus (Fig. 1.21). The cerebral and cerebellar falces are like sickles that split the cerebrum (Fig. 1.22) and cerebellum (Fig. 1.48) into bilateral hemispheres (Fig. 1.20).

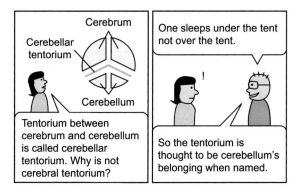

Fig. 1.20

The cerebellar tentorium (Fig. 1.19) is like a tent of the cerebellum.

Blood in the cerebral "vein" drains to the dural "venous" sinus (e.g., superior sagittal sinus). The dural venous sinus also receives cerebrospinal fluid from the subarachnoid space via arachnoid granulation, which is extension of the arachnoid mater (Fig. 1.17).

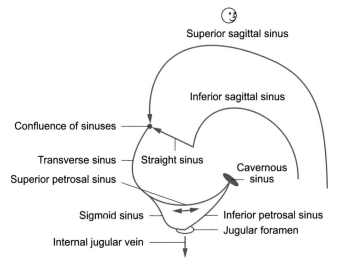

Fig. 1.21 Dural venous sinuses.

The above figure shows the direction of blood flow (including cerebrospinal fluid) through the dural venous sinuses. Eventually, all blood empties into the internal jugular vein.

This subchapter has explored the blood circulation from the vertebral and internal carotid arteries (Fig. 1.3) to the internal jugular vein. These arteries and vein occupy quite different locations from one another.

Morphology of the cerebral hemisphere

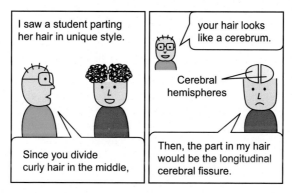

Fig. 1.22

The cerebrum consists of two cerebral hemispheres (Figs. 1.5, 1.11).

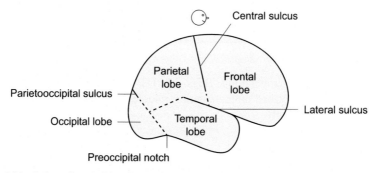

Fig. 1.23   Lobes of cerebral hemisphere.

On the lateral surface of a cerebral hemisphere, the most and the second most deep sulci are the lateral and central sulci. They are the borders between the frontal, parietal, and temporal lobes. The occipital lobe is demarcated by the less distinct parietooccipital sulcus (Fig. 1.28) and preoccipital notch.

14

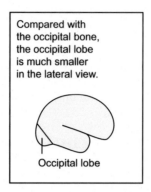

Compared with the occipital bone, the occipital lobe is much smaller in the lateral view.

Occipital lobe

Fig. 1.24

The four lobes roughly correspond to the frontal, parietal, temporal, and occipital bones of the skull.

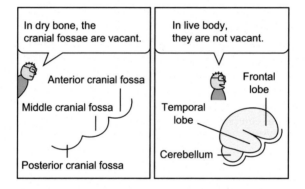

In dry bone, the cranial fossae are vacant.

Anterior cranial fossa

Middle cranial fossa

Posterior cranial fossa

In live body, they are not vacant.

Frontal lobe

Temporal lobe

Cerebellum

Fig. 1.25

The frontal lobe, temporal lobe (Figs. 1.5, 1.23), and cerebellum (with brainstem) (Fig. 1.2) are placed on the anterior, middle, and posterior cranial fossae of the skull, respectively.

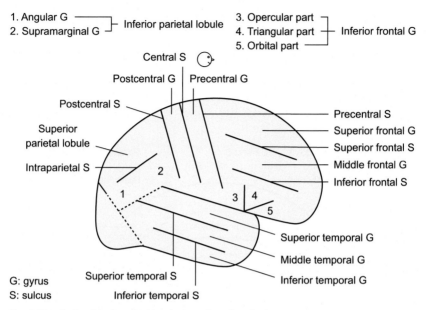

Fig. 1.26   Gyri, sulci of cerebral hemisphere (lateral surface).

Each lobe consists of gyri, bordered by sulci (Fig. 1.31). For example, the frontal lobe (lateral surface) consists of the superior, middle, and inferior frontal gyri, bordered by the superior and inferior frontal sulci, excluding the precentral gyrus. The inferior frontal gyrus is subdivided into the opercular part, triangular part (Fig. 4.12), and orbital part.

The inferior parietal lobule is composed of the angular and supramarginal gyri. The "angular" gyrus, encountering the superior temporal sulcus, occupies an "angle" of the parietal lobe, surrounded by the occipital and temporal lobes. The "supramarginal" gyrus is "above margin" of the lateral sulcus.

The lateral surface does not show the transverse temporal gyrus, which is the floor of lateral sulcus (Fig. 1.40).

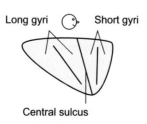

Fig. 1.27   Insula.

16

Around the lateral sulcus, the frontal, parietal, and temporal lobes hide the insula, an independent lobe (Fig. 1.40). The "opercular" part of the inferior frontal gyrus is an "operculum" (lid) of the insula (Fig. 1.26).

The insula is made up of the short gyri and long gyri, on either side of its central sulcus. The central sulcus of insula has the same name as that between the frontal and parietal lobes (Fig. 1.23). This is because the two central sulci are on the same oblique plane.

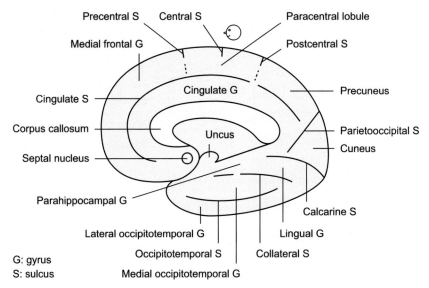

Fig. 1.28   Gyri, sulci of cerebral hemisphere (medial surface).

Gyri and sulci can also be found on the medial surface of the cerebral hemisphere (viewed from the longitudinal cerebral fissure) (Fig. 1.5). On the medial surface, the parietooccipital sulcus is distinct, while the other borders between the four lobes are indistinct (Fig. 1.23).

The precentral and postcentral gyri are connected at the medial surface by the paracentral lobule (Fig. 4.8). There is a tendency to use the term "lobule" when two gyri are grouped. Another example is the inferior parietal lobule comprising the angular and supramarginal gyri. An exception is the superior parietal lobule (Fig. 1.26) which is continuous with only one gyrus, the precuneus.

The cingulate sulcus *(exactly, cingulate sulcus and subparietal sulcus)* surrounds the cingulate gyrus. The arch-shaped cingulate gyrus surrounds the corpus callosum, which is the main commissure of the bilateral cerebral hemispheres (Fig. 1.40). The cingulate gyrus is connected with the parahippocampal gyrus morphologically and functionally (Fig. 4.13). The anterior part of parahippocampal gyrus curves backward as the uncus (meaning hook).

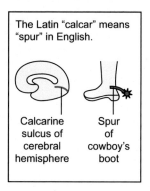

Fig. 1.29

The calcarine sulcus at the posterior end of the cerebral hemisphere (Fig. 1.28) is similar in location and shape to the spur of cowboy's boot.

The calcarine sulcus and parietooccipital sulcus border the cuneus, which is a medial gyrus of the occipital lobe (Fig. 1.23). The calcarine sulcus and collateral sulcus border the lingual gyrus, which is an inferior gyrus of the occipital and temporal lobes (Fig. 3.10). The cuneus looks like a wedge, whereas the lingual gyrus looks like a long tongue (Fig. 1.28).

The occipitotemporal sulcus runs between the medial and lateral occipitotemporal gyri (Fig. 1.28). Sometimes, the medial occipitotemporal gyrus is called the fusiform gyrus; the lateral occipitotemporal gyrus is regarded as a part of the inferior temporal gyrus (Figs. 1.26, 1.30).

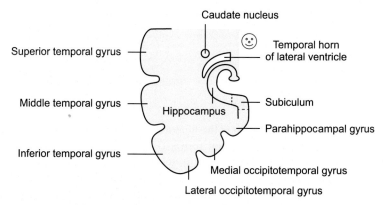

Fig. 1.30 Temporal lobe (coronal plane).

To put it concretely, the parahippocampal, medial occipitotemporal, and lateral occipitotemporal gyri make up inferomedial surface of the temporal lobe (Fig. 1.28).

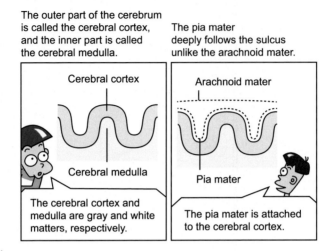

Fig. 1.31

Each gyrus consists of the cerebral cortex and cerebral medulla (Fig. 4.1). When the cerebrum is cut, the cerebral cortex (gray color) and cerebral medulla (white color) are easily distinguishable (Fig. 5.10). The color difference can be recognized in brain MRI too. (T1-weighted MRI displays the gray matter in gray color and the white matter in white color.)

Morphology of the limbic system

The cerebrum includes the "limbic" system that forms a "limbus" in the cerebrum (Fig. 4.13).

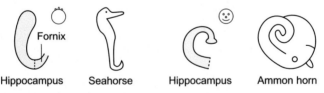

Fig. 1.32 Hippocampus, similar things.

The center of limbic system is the hippocampus, which is a primitive cerebral cortex. The hippocampus and fornix (Fig. 1.35) in the superior view resemble a seahorse. A seahorse lives in water; but the HIPPOcampus (like a HIPPOpotamus) lives near water which is the cerebrospinal fluid in the temporal horn of lateral ventricle (Fig. 1.30). The hippocampus in the coronal plane is shaped like Ammon horn. (According to other assertion, the hippocampus in the coronal plane also resembles a seahorse.)

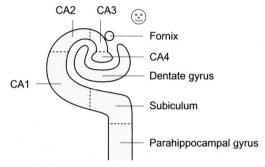

Fig. 1.33 Hippocampus, adjacent structures (coronal plane).

Histologically, the hippocampus is subdivided into CA1, CA2, CA3, CA4 [CA = Cornu Ammonis = Ammon horn (Fig. 1.32)]. The parahippocampal gyrus (a cerebral cortex) (Fig. 1.28) is connected to the subiculum, CA1, CA2, CA3, CA4 in sequence.

The hippocampus is in contact with the fornix (a bundle of axons, like the tract). *[Exactly, it is the fimbria of hippocampus; the fimbria becomes the fornix after the axons leave the hippocampus area (Fig. 1.35).]* The dentate gyrus, which is another primitive cerebral cortex, is facing the hippocampus.

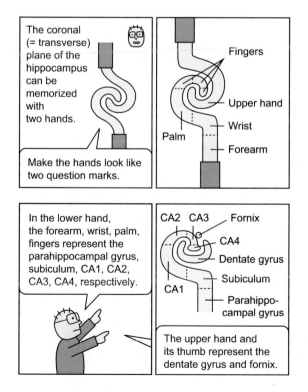

Fig. 1.34

Two hands held with the fingers wrapping each other resemble the hippocampus and dentate gyrus.

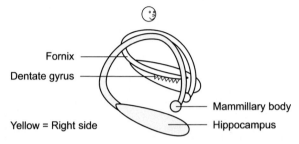

Fornix
Dentate gyrus
Yellow = Right side
Mammillary body
Hippocampus

Fig. 1.35   Hippocampus, fornix.

The above figure shows that the dentate gyrus and fornix are located on the medial side of the hippocampus. The fornix is an arch extending from the hippocampus (Fig. 1.33) to the mammillary body, which is a part of hypothalamus (Figs. 1.44, 1.62).

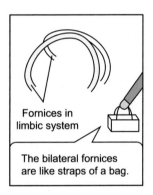

Fornices in limbic system

The bilateral fornices are like straps of a bag.

Fig. 1.36

The bilateral fornices meet each other (partly decussate), being observable in the median plane (Figs. 1.42, 1.44).

Morphology of the basal nuclei

Nerve cell bodies (Fig. 2.2) situated in the peripheral nervous system are called ganglia, while those located in the central nervous system are called nuclei (Fig. 2.8). Thus the commonly used term "basal ganglia" needs to be fixed to the term "basal nuclei."

```
┌─────────────────────────────────────────────────────────────┐
│  Corpus striatum ┌─ Striatum — *Putamen, caudate nucleus     │
│                  └─ Pallidum — *Globus pallidus              │
│  Subthalamus                                                 │
│  Substantia nigra              *Lentiform nucleus            │
└─────────────────────────────────────────────────────────────┘
```

Fig. 1.37   Composition of basal nuclei.

The corpus striatum, subthalamus, and substantia nigra are all basal nuclei. The corpus striatum is divided into the striatum (putamen and caudate nucleus) and pallidum (globus pallidus) (Fig. 4.18).

Fig. 1.38   Projection of corpus striatum.

The corpus striatum, the center of "basal" nuclei, is located at the "basal" area of the cerebrum.

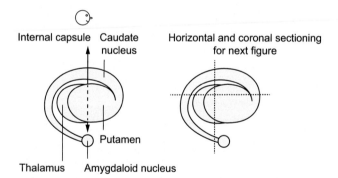

Fig. 1.39   Corpus striatum, adjacent structures (lateral view).

During development, the caudate nucleus is elongated to become C-shaped (270 degrees angle). This elongation determines the shape of both the lateral ventricle (Fig. 1.12) and cerebrum (Fig. 5.12). The elongated caudate nucleus tapers and reaches the amygdaloid nucleus that does not belong to basal nuclei (Fig. 1.37), but to the limbic system (Fig. 4.14).

The term "caudate" means "tail shape" while "caudal" [e.g., caudal medulla oblongata (Fig. 1.11)] means "tail direction" (Fig. 5.5).

22

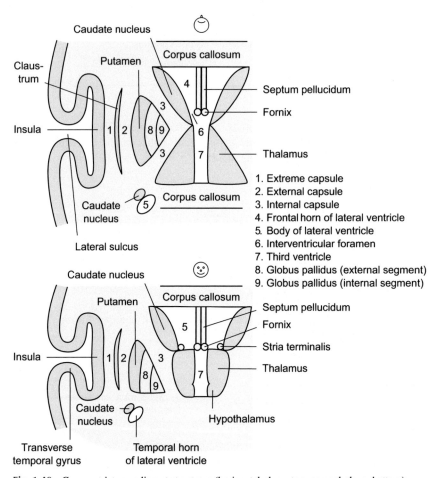

Fig. 1.40 Corpus striatum, adjacent structures (horizontal plane, top; coronal plane, bottom).

1. Extreme capsule
2. External capsule
3. Internal capsule
4. Frontal horn of lateral ventricle
5. Body of lateral ventricle
6. Interventricular foramen
7. Third ventricle
8. Globus pallidus (external segment)
9. Globus pallidus (internal segment)

The C-shaped caudate nucleus can be seen twice in the horizontal plane and also in the coronal plane (Fig. 1.39). Every portion of the caudate nucleus is in contact with the lateral ventricle (Fig. 1.41).

In the coronal plane, the caudate nucleus forms the lateral wall of the body of lateral ventricle (Fig. 1.41); the thalamus forms its floor. Between the lateral wall and floor, the stria terminalis runs from the amygdaloid nucleus along with the caudate nucleus (Figs. 1.39, 4.14). The stria terminalis (meaning "boundary" tract) forms "boundary" between the caudate nucleus and thalamus.

In the horizontal plane, the two lateral "ventricles" are connected with the third "ventricle" via the "interventricular" foramina (Fig. 1.42). In the coronal plane, the third ventricle is located between the bilateral thalami and hypothalami (Figs. 1.11, 1.41, 1.44).

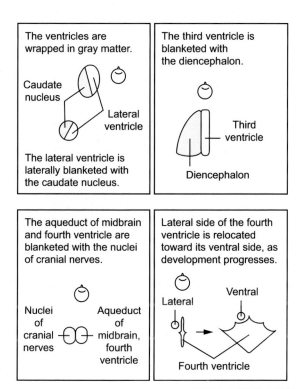

Fig. 1.41

The ventricles are covered by gray matter on its lateral side: the lateral ventricle by the caudate nucleus, the third ventricle by the diencephalon (Fig. 1.40), and the aqueduct of midbrain and fourth ventricle by the nuclei of cranial nerves (Figs. 5.18, 5.21, 5.22). In the initial form, the neural canal is covered by the intermediate zone (gray matter) on its lateral side (Fig. 5.14).

In the horizontal and coronal planes (Fig. 1.40), the putamen and the globus pallidus (external and internal segments) form the LENtiform nucleus which is LENs-shaped (Fig. 1.37). The GLOBus PALidus is a GLOBe which is PALer than the putamen due to the larger amount of myelin sheaths (Fig. 5.10). Nevertheless, the globus pallidus is a gray matter, including nerve cell bodies like the putamen (Figs. 4.16, 4.18), so the globus pallidus is darker than the white matter such as the adjacent internal capsule.

The two planes (Fig. 1.40) also show the cerebral cortex (gray matter) and cerebral medulla (white matter). The cerebral medulla includes the internal, external, and extreme capsules, which carry ascending and descending axons of sensory and motor nerves (Figs. 2.8, 2.17). The most prominent one is the internal capsule (Fig. 5.11).

The "internal" capsule is "internal" to the lentiform nucleus; the "external" capsule is "external" to the lentiform nucleus. They look like the capsule of the lentiform nucleus (Fig. 1.40).

The "internal" capsule is bent to the "internal" direction in the horizontal plane. More internal to the internal capsule, there exist two other gray matter structures: the caudate nucleus and thalamus (Figs. 1.39, 5.11). Overall, the Lentiform nucleus is Lateral to the meaningful structures (Fig. 1.40).

The extreme capsule is between the claustrum and insula (Figs. 1.27, 1.40). The insula is covered by the frontal, parietal, and temporal lobes (in detail, their opercula) (Fig. 1.26).

The septum pellucidum (meaning translucent septum) is actually opaque in cadaver specimen. The septum pellucidum runs from the corpus callosum down to the fornix (Figs. 1.40, 1.42, 1.44).

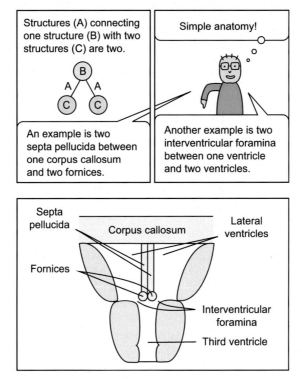

Fig. 1.42

The septa pellucida exist in tandem like the fornices. Two septa pellucida are in contact because the two fornices are in contact (Fig. 1.35); but septa pellucida can be separated by meticulous dissection. The septa pellucida are the septa between the lateral ventricles (frontal horns and bodies) (Figs. 1.12, 1.40, 1.44).

Morphology of the diencephalon

Whereas the corpus striatum (lentiform nucleus and caudate nucleus) is gray matter within the cerebrum (Fig. 1.38), the diencephalon is gray matter between the cerebrum and brainstem (Fig. 1.11).

The diencephalon cannot be seen from the side due to the drastic growth of the cerebral hemisphere (Fig. 1.11), so the diencephalon is occasionally disregarded when speaking about the main brain components (cerebrum, cerebellum, and brainstem) (Fig. 1.2). However, the diencephalon is indeed an independent component (Fig. 5.6).

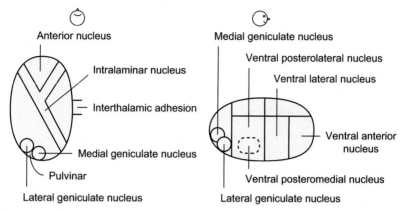

Fig. 1.43   Thalamic nuclei.

As the main part of the diencephalon, the thalamus comprises plenty of nuclei. Pulvinar is the posterior part of the thalamus (Fig. 1.45) above the medial and lateral geniculate nuclei.

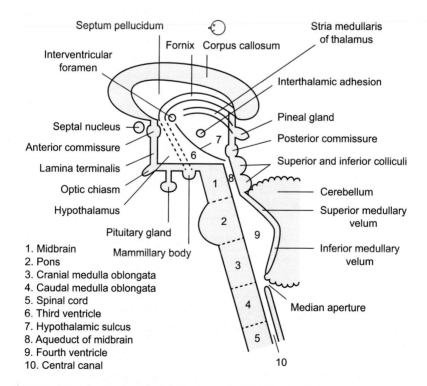

1. Midbrain
2. Pons
3. Cranial medulla oblongata
4. Caudal medulla oblongata
5. Spinal cord
6. Third ventricle
7. Hypothalamic sulcus
8. Aqueduct of midbrain
9. Fourth ventricle
10. Central canal

Fig. 1.44   Diencephalon (medial view), brainstem (median plane), adjacent structures.

The interthalamic adhesion connects the bilateral thalami (Figs. 1.43, 5.17). The interthalamic adhesion, which has no particular function, has a landmark role in neuroanatomy and neuroimaging.

The other parts of diencephalon are the epithalamus, hypothalamus (Fig. 5.17), and subthalamus (Fig. 4.26).

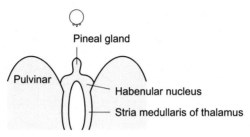

Fig. 1.45 Epithalamus.

Superoposterior to the thalamus, there is the epithalamus which consists of the stria medullaris of thalamus, habenular nucleus, and pineal gland (Fig. 1.44). The "stria medullaris of thalamus" is a tract which is implied by "stria"; it is an inner part which is implied by "medullaris"; it belongs to the epithalamus in spite of the modifier "of thalamus."

Functionally, the stria medullaris of thalamus and habenular nucleus are influenced by the limbic system (Fig. 4.14); the PINEal gland that looks like a PINE cone secretes hormone (melatonin), so it also belongs to the endocrine system.

Inferoanterior to the thalamus is the hypothalamus, which is independent gray matter. The border between the thalamus including interthalamic adhesion (Fig. 1.43) and the hypothalamus including mammillary body is the hypothalamic sulcus (Figs. 1.44, 4.26, 5.16, 5.17).

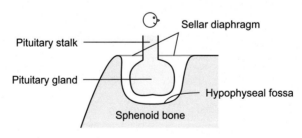

Fig. 1.46 Pituitary gland.

The pituitary gland (Fig. 1.62), a part of the endocrine system, is suspended from the hypothalamus by the pituitary stalk (Figs. 1.44, 4.27). The pituitary gland, namely "hypophysis" is cradled within the "hypophyseal" fossa of the sphenoid bone. The sellar diaphragm covers the pituitary gland from above (Fig. 1.19).

The pituitary gland is about the size of a pea, but is divided into two segments: neurohypophysis and adenohypophysis (Figs. 4.27, 5.13).

The spindle-shaped subthalamus is located lateral to the hypothalamus (Fig. 4.26), so the subthalamus is invisible in the medial view of the diencephalon (Fig. 1.44). In terms of function, the subthalamus belongs to the basal nuclei (Figs. 1.37, 4.18).

In Fig. 1.44, structures connecting the bilateral cerebral hemispheres are discernible: the lamina terminalis (Fig. 1.11), the anterior commissure, the corpus callosum which is the largest one (Fig. 1.40), and the posterior commissure. The commissural neuron passes through these structures (Fig. 4.5).

Around the anterior commissure and corpus callosum is the septal nucleus (Figs. 4.13, 4.14), which is a part of the frontal lobe (Fig. 1.28). The "septal" nucleus is under the "septum" pellucidum (Fig. 1.44).

Morphology of the cerebellum

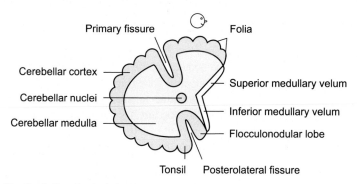

Fig. 1.47   Cerebellum (sagittal plane).

On the cerebellum are numerous folds that enlarge the cerebellar cortex like the gyri of cerebrum (Fig. 5.9). The folds on the cerebellum are too small to be called gyri, and so are referred to as folia (meaning leaves). On the surface of the cerebellum, the folia show leaf pattern. *[Exactly, each folium involves both the cerebellar cortex and cerebellar medulla (Fig. 1.31) unlike the above figure.]*

Between the numerous folia, there are two distinct fissures: the posterolateral fissure and the primary fissure. The posterolateral fissure is the posterior boundary of flocculonodular lobe (Fig. 1.48). The flocculonodular lobe is evolutionally old (Fig. 4.34); therefore, the posterolateral fissure is older than the primary fissure. The word "primary" does not mean evolutional oldness but morphological deepness.

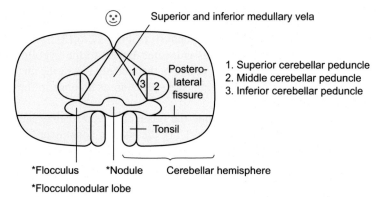

Fig. 1.48 Cerebellum (ventral view).

The superior, middle, and inferior cerebellar peduncles are made of white matter that connects the cerebellum to the midbrain, pons, and medulla oblongata, respectively (Figs. 1.57, 4.38). After cutting the cerebellar peduncles of a cadaver, the cerebellum can be detached from the brainstem. The detached cerebellum contains the superior and inferior medullary vela which are the roof of fourth ventricle (Figs. 1.44, 1.47, 1.54, 1.58).

In the ventral view of the detached cerebellum, the flocculonodular lobe, composed of two flocculi and one nodule, can be identified. The "posterolateral" fissure is "posterior" to the flocculonodular lobe (Fig. 1.47), and "lateral" to it.

The tonsils are paired like the palatine tonsils are paired in fauces. The tonsils, caudal to the flocculonodular lobe, are the most caudal portion of the cerebellum (Fig. 1.47). Therefore, in case of extremely high intracranial pressure, the tonsils may herniate through the foramen magnum (Fig. 3.53) and press the medulla oblongata (Fig. 1.44) to cause fatal results (Fig. 4.28).

Just like the cerebrum (Fig. 1.31), the cerebellum is divided into the cerebellar cortex (gray matter), cerebellar medulla (white matter), and cerebellar nuclei (gray matter) (Fig. 1.47). The cerebellar nuclei among the cerebellar medulla are equivalent to the basal nuclei among the cerebral medulla (Figs. 1.38, 1.40, 4.39).

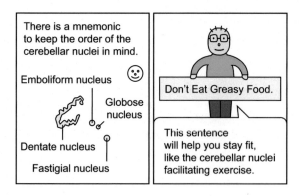

Fig. 1.49

From lateral to medial, the four cerebellar nuclei are the dentate, emboliform, globose, and fastigial nuclei (Fig. 1.50).

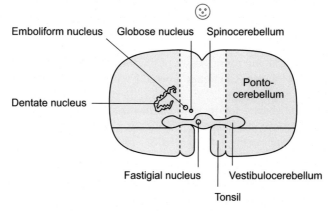

Fig. 1.50 Vestibulocerebellum, spinocerebellum, pontocerebellum.

There are three cerebella, divided according to their functions (Figs. 4.34, 4.35, 4.36). The smallest vestibulocerebellum is the flocculonodular lobe (Fig. 1.47); it is related with the fastigial nucleus. The spinocerebellum occupies the medial area of the cerebellum; it is related with the emboliform and globose nuclei. The biggest pontocerebellum occupies the "lateral" area, so it is connected with the pons by way of the middle cerebellar peduncle which is "lateral" (Figs. 1.48, 1.54). The pontocerebellum is related with the macroscopic dentate nucleus.

Morphology of the brainstem

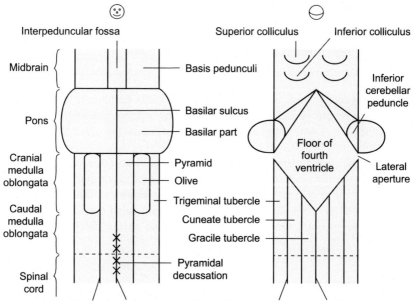

Fig. 1.51 Brainstem (ventral view, left; dorsal view, right).

Four parts of the brainstem are the midbrain containing the aqueduct of midbrain, the pons and cranial medulla oblongata containing the fourth ventricle, and the caudal medulla oblongata containing the central canal (Figs. 1.11, 1.44). Ventral and dorsal views of the four parts will be compared with their transverse planes. Challenge yourself with the game of stereoscopic recognition.

The pyramid-shaped fourth ventricle has the diamond-shaped floor (Fig. 5.19). Such stereoscopic shape is reflected in the median plane (Fig. 1.44) and transverse planes (Figs. 1.54, 1.58).

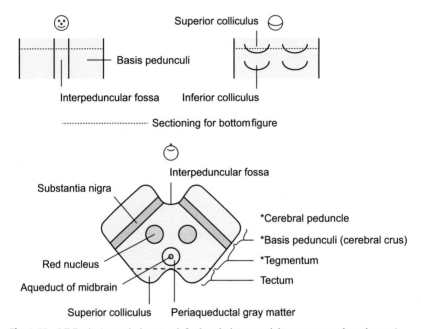

Fig. 1.52   Midbrain (ventral view, top left; dorsal view, top right; transverse plane, bottom).

In the ventral view of the midbrain, the basis pedunculi and interpeduncular fossa are visible; in its dorsal view, the superior and inferior colliculi are visible (Fig. 1.44). These can be seen in the transverse plane as well (Fig. 3.23).

In the transverse plane, the midbrain is divided into three parts: the basis pedunculi, the tegmentum (including the substantia nigra, red nucleus, and periaqueductal gray matter), and the tectum (consisting of the superior and inferior colliculi) (Fig. 1.44). The basis pedunculi (also called the cerebral crus) and tegmentum are collectively referred to as the cerebral peduncle (Fig. 1.57).

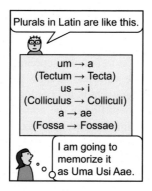

Fig. 1.53

Three plural forms in Latin are frequently used in neuroanatomy.

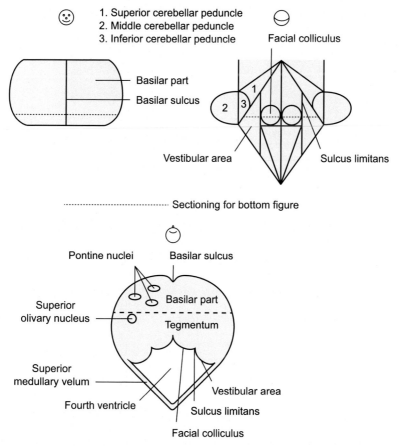

Fig. 1.54 Pons (ventral view, top left; dorsal view, top right; transverse plane, bottom).

In the ventral view and transverse plane of the pons, the basilar part and basilar sulcus are visible. On the "basilar" sulcus, the "basilar" artery is accommodated (Fig. 1.3).

In the dorsal view of above figure, the superior medullary velum and cerebellum (Figs. 1.44, 1.48) are removed to show the floor of fourth ventricle (Fig. 1.51). The sulcus limitans that is the border between the vestibular area and the facial colliculus is identified in the dorsal view and transverse plane (Fig. 5.21). The facial "colliculus" is a bulging area because of the inside nerve cell bodies (Fig. 3.25) like the superior "colliculus" (Figs. 1.52, 4.45, 4.46) and inferior "colliculus" (Fig. 3.52).

The transverse plane shows that the pons is divided into the basilar part (including the pontine nuclei) and tegmentum (including the superior olivary nucleus). The tegmentum is like a ceiling, while the tectum is like a roof (Fig. 1.52). The pons is a house with a ceiling, without a roof.

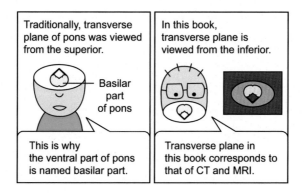

Fig. 1.55

The terms "basis pedunculi, basilar part, tegmentum, and tectum" (Figs. 1.52, 1.54) were coined when the transverse plane was conventionally viewed from the superior. However, this book employs the different transverse plane, where the ventral side of a structure is at the top of the figure.

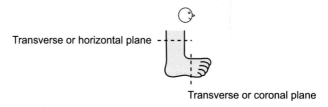

Fig. 1.56   Transverse plane.

The transverse (cross) plane, which is at a right angle to the axis, is not always the same as the horizontal plane. For example, the transverse plane of a leg is the horizontal plane, while the transverse plane of a foot is the coronal plane (Fig. 1.39). In neuroanatomy, the transverse plane is more useful than the horizontal plane because of the varying flexion angles of the neural tube (axis) (Fig. 5.5).

The pons is the thickest part of the brainstem (Figs. 1.44, 1.51) because of the big pontine nuclei in the basilar part (Fig. 1.54).

The cerebral peduncle of midbrain connects the cerebrum with the brainstem (mainly, pons) (Fig. 1.52), while the cerebellar peduncle connects the cerebellum with the brainstem (mainly, pons) (Fig. 1.54). A great deal of impulses from the cerebrum go to the cerebellum (Figs. 4.37, 4.39) by way of the cerebral peduncle, pons, and middle cerebellar peduncle. Embryologically, both the pons and cerebellum originate from the metencephalon (Fig. 5.6).

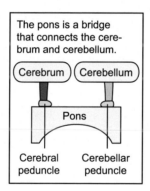

Fig. 1.57

Etymologically, the PEDuncle is foot like the PEDestrian is traveler on foot; the PONs is bridge like PONt Neuf is bridge in Paris. The cerebrum and cerebellum set their feet on a bridge (pons).

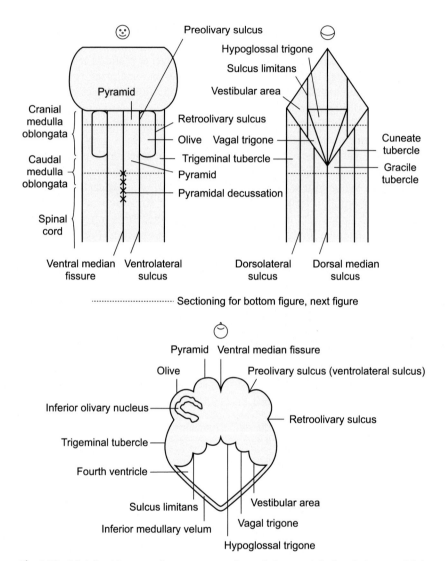

Fig. 1.58 Medulla oblongata, adjacent structures (ventral view, top left; dorsal view, top right), cranial medulla oblongata (transverse plane, bottom).

The caudal part of the brainstem is the medulla oblongata, which is regarded as the spinal cord (medulla) that is elongated (oblongata). The medulla oblongata has spinal cord structures such as the ventral median fissure, ventrolateral sulcus, dorsolateral sulcus, and dorsal median sulcus (Figs. 1.59, 1.68).

The term "medulla" means not only the spinal cord [e.g., conus medullaris (Fig. 1.66)] but also the inner part [e.g., cerebral medulla (Fig. 1.31)].

Regarding the fissure and sulci of the medulla oblongata as boundaries, there exist the pyramid, olive, trigeminal tubercle, cuneate tubercle, and gracile tubercle (Fig. 1.59).

The pyramid is visible both in the cranial and caudal medulla oblongata (Fig. 1.59), but the "olive" is visible only in the cranial medulla oblongata where the inside structure (inferior "olivary" nucleus) is present. The inferior olivary nucleus is morphologically similar to the dentate nucleus, both of which are medially concave (Fig. 1.50). Meanwhile, the superior olivary nucleus can be seen in the tegmentum of pons (Fig. 1.54).

Let's focus on the cranial medulla oblongata (ventral view). The cranial part of the ventrolateral sulcus is referred to as the preolivary sulcus. The next sulcus between the olive and trigeminal tubercle is the retroolivary sulcus. (Official terms are the preolivary and retroolivary grooves.)

The pyramid, olive, trigeminal tubercle and their demarcating fissure, sulci are identifiable in the transverse plane.

The floor of fourth ventricle illustrates the sulcus limitans between the vestibular area and the vagal, hypoglossal trigones. The related cranial nerves (CN VIII, X, XII) are in arithmetic progression with a common difference of 2. The three structures can be observed in the transverse plane as well (Fig. 5.22). The anatomy term "trigone" is used for the triangular area that is slightly swollen, as the trigone of bladder.

As the superior medullary velum is the roof of fourth ventricle in the pons (Fig. 1.54), the inferior medullary velum is that in the cranial medulla oblongata (Fig. 1.44).

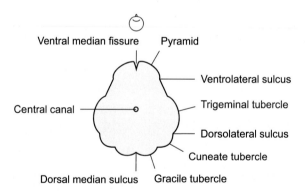

Fig. 1.59   Caudal medulla oblongata (transverse plane).

The circular transverse plane of the caudal medulla oblongata seems to be the origin of the medulla oblongata's another name "bulb" (e.g., corticobulbar tract). The caudal medulla oblongata contains the central canal (Fig. 1.11).

Compare this transverse plane with the ventral and dorsal views of the caudal medulla oblongata (Fig. 1.58 top). In the transverse plane, external features (pyramid, trigeminal tubercle, cuneate tubercle, gracile tubercle, and related fissure, sulci) are recognizable.

The pyramidal decussation between the bilateral pyramids extends to the uppermost part of spinal cord (Figs. 1.51, 2.17, 2.19). The pyramidal decussation makes the ventral median fissure shallow (Fig. 1.68).

36

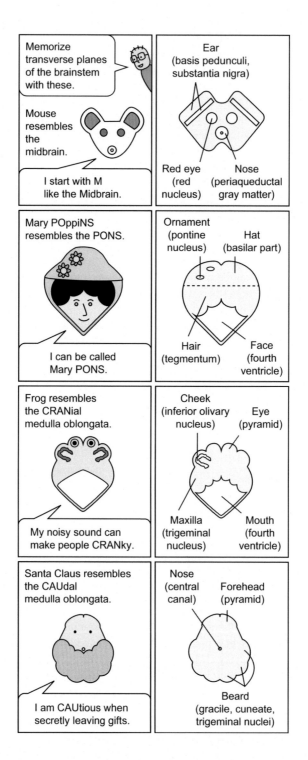

Fig. 1.60

The four transverse planes of the brainstem should be memorized by any means.

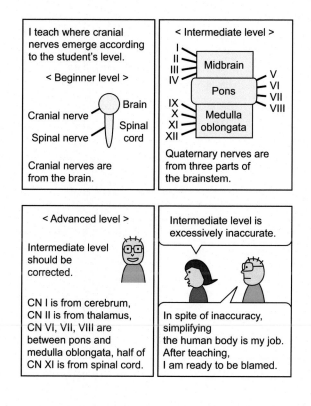

Fig. 1.61

Cranial nerves emerge from the brainstem. Exceptions are CN I from the cerebrum, CN II from the thalamus (Figs. 3.1, 3.5, 3.83), and spinal root of CN XI from the spinal cord (Figs. 1.68, 3.53, 3.64).

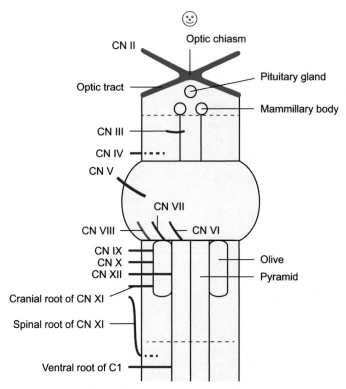

Fig. 1.62 Cranial nerves emerging from brainstem.

CN II, the pituitary gland, and the mammillary body are not in the brainstem, but in the diencephalon. The Pituitary gland is Posterior to the optic chiasm; the Mammillary body is the Most posterior among these three structures (Fig. 1.44).

CN III is from the interpeduncular fossa (Fig. 3.12); CN IV is from the dorsal surface below the inferior colliculus (Fig. 3.24); CN V is from the basilar part of pons (Fig. 1.54); CN VI, VII, VIII are from the border between pons and medulla oblongata; CN IX, X, cranial root of CN XI are from the retroolivary sulcus of cranial medulla oblongata (Fig. 1.58); spinal root of CN XI is from the spinal cord (Figs. 1.68, 3.53, 3.64); and CN XII is from the preolivary sulcus of cranial medulla oblongata (Fig. 3.66).

12 pairs of cranial nerves
exit the cranial cavity.

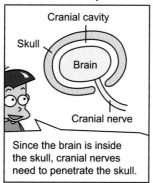

Fig. 1.63

All cranial nerves pass through foramina or canals to exit the cranial cavity. For instance, CN IX, X, XI pass through the jugular foramen (Fig. 3.53).

Speaking of the peripheral
nervous system, cranial
nerves are mostly in the head,
while spinal nerves are
in the trunk and limbs.

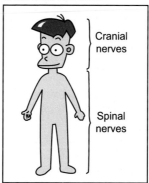

Fig. 1.64

Cranial nerves are distributed to the head and to a certain part of the neck. An exception is CN X, which is also distributed to the thoracic and abdominal cavities (Figs. 2.32, 2.33, 2.34).

Morphology of the spinal cord

Some people play a prank by pulling a chair.

However, this could lead to a severe injury of the spinal cord.

If the spinal cord is disconnected, the impulse will not be transmitted below the disconnection site.

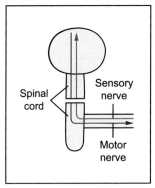

Fig. 1.65

Unlike the peripheral nervous system, the central nervous system cannot be repaired after disconnection (Fig. 2.10). That is why the brain and spinal cord are so preciously protected by the skull (Fig. 1.63) and vertebral column (Fig. 1.8).

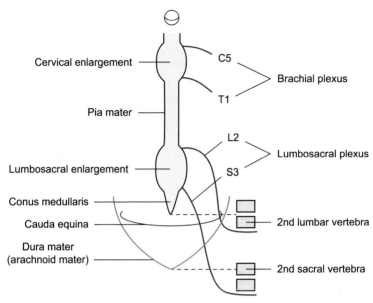

Fig. 1.66   Spinal cord, adjacent structures.

The spinal dura mater ends at the 2nd sacral vertebra. We use the 2nd sacral vertebra to describe the level, even though the 1st—5th sacral vertebrae (Fig. 3.70) fuse to form the sacrum (Fig. 2.35) during adolescence. The arachnoid mater (subarachnoid space) ends at the same level because the subdural space is a potential space (Fig. 1.8).

The inferior part of the spinal cord is the conus medullaris (cone of spinal cord). At a very early stage of embryological development, the vertebral column used to be as long as the spinal cord. But after this stage, vertical growth of the vertebral column is faster than that of the spinal cord. Because of this discrepancy in length, the conus medullaris ends at the level between the 1st and 2nd lumbar vertebrae in adults (at the level between the 2nd and 3rd lumbar vertebrae in newborns).

Lower spinal nerves from the spinal cord are longer, to reach the corresponding intervertebral foramina. For example, L2 reaches the intervertebral foramen between the 2nd and 3rd lumbar vertebrae (Fig. 3.70). These spinal nerves below the conus medullaris are named the cauda equina (tail of horse), due to their resemblance of appearance.

42

There are two enlarged parts in the spinal cord, known as the cervical and lumbosacral enlargements.

Spinal nerves that extend to the upper and lower limbs emerge from the cervical and lumbosacral enlargements, respectively.

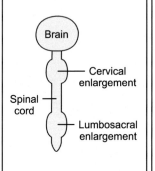

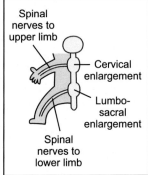

Fig. 1.67

The cervical enlargement of the spinal cord is for the brachial plexus (C5−T1), innervating the upper limb (Fig. 3.78). Likewise, the "lumbosacral" enlargement is for the "lumbosacral" plexus (L2−S3), innervating the lower limb (Figs. 1.66, 3.81). *(Exactly, T11−S1 emerge from the lumbosacral enlargement.)*

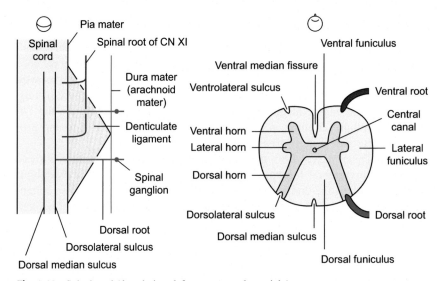

Fig. 1.68 Spinal cord (dorsal view, left; transverse plane, right).

The spinal cord is stabilized by the denticulate ligaments extending laterally to the dura mater (Fig. 3.64). The "denticulate" ligaments, the "dentate" gyrus in limbic system (Fig. 1.35), and the "dentate" nucleus in cerebellum (Fig. 1.50) altogether look like sharp teeth.

The ventral median fissure is much deeper than the dorsal median, ventrolateral, dorsolateral sulci. A fissure is a deep sulcus. In the cerebrum, the longitudinal cerebral fissure is notably deep (Fig. 1.5). In the cerebellum, the posterolateral and primary fissures are deep (Fig. 1.47).

In the transverse plane of a cadaver's spinal cord, the central gray matter and the peripheral white matter can be distinguished with the naked eye. Regarding the neurons, the former contains mainly the nerve cell bodies, while the latter contains only the axons and myelin sheaths (Figs. 2.24, 5.10). This structure of the spinal cord is close to the original form of the neural tube (Fig. 5.8).

Due to the ventral and dorsal horns, the gray matter looks like H beam of the spinal cord morphologically. The ventral and dorsal horns of gray matter correspond to the ventrolateral and dorsolateral sulci (and then ventral and dorsal roots).

In addition, the lateral horn is visible at the level of T1−L2, which involves the sympathetic nerve (Fig. 2.28). The "central" canal (Fig. 1.11) is located in the "center" of the "central" gray matter.

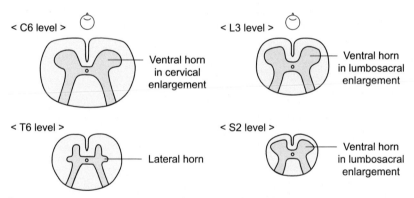

Fig. 1.69   Gray and white matters of spinal cord (transverse planes).

The cervical and lumbosacral enlargements (Fig. 1.66) are the result of the increased volume of the ventral horn. The ventral horn is full of the numerous lower motor neurons (nerve cell bodies) of the corticospinal tract (Fig. 2.19), which innervate the big muscles of the upper and lower limbs (Fig. 1.67). At these levels, the dorsal horn is relatively thick due to many 2nd neurons (nerve cell bodies) of spinothalamic tract (Fig. 2.11) and dorsal, ventral spinocerebellar tract (Figs. 4.36, 4.38).

The white matter is roughly divided into the ventral funiculus, dorsal funiculus, and lateral funiculus (Fig. 1.68). The volume of white matter increases in proportion to the level of spinal cord. Both the sensory and motor nerves passing longitudinally in the cranial spinal cord are thicker than those in the caudal spinal cord (Fig. 2.14). It is like the proximal water pipe being thicker than the distal one.

Chapter 2

The somatic nerve, the autonomic nerve

The nervous system consists of four kinds of nerves: the somatic sensory nerve from receptor around the skeletal muscle (e.g., receptor in the skin), the somatic motor nerve to the skeletal muscle (voluntary muscle), the visceral sensory nerve from the receptor around the smooth or cardiac muscle (e.g., receptor in the gastrointestinal tract), and the visceral motor nerve (autonomic nerve) to the smooth or cardiac muscle (involuntary muscle). This chapter explores two pathways of the somatic sensory nerve (spinothalamic tract and medial lemniscus pathway), one pathway of the somatic motor nerve (corticospinal tract). Additional content is two components of the visceral motor nerve (sympathetic and parasympathetic nerves). Comprehension of the pathways is enhanced by practice with stained slices of the brainstem and spinal cord (or their photos).

The neuron

Prior to the somatic nerve and autonomic nerve, the neuron will be introduced as orientation.

The nervous system consists of innumerable neurons which transmit impulse

and neuroglias which support the neurons.

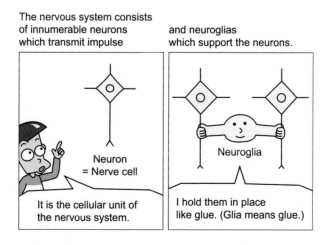

Neuron
= Nerve cell

It is the cellular unit of the nervous system.

Neuroglia

I hold them in place like glue. (Glia means glue.)

Fig. 2.1

Visually Memorable Neuroanatomy for Beginners. DOI: https://doi.org/10.1016/B978-0-12-819901-5.00002-8

Note that neuron is synonymous with nerve cell. The neuron is physically and functionally supported by neuroglia.

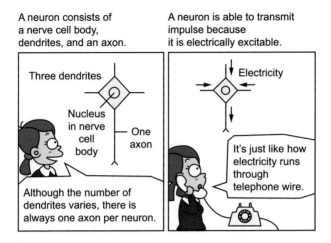

Fig. 2.2

A typical neuron, multipolar neuron, has two or more dendrites and one axon. The dendrites convey impulse to the nerve cell body, while the axon conveys impulse from the nerve cell body.

Each nerve cell body contains a nucleus full of chromosomes. The aggregated nerve cell bodies in the brain are also called a nucleus (Figs. 1.43, 2.8). The term "nucleus" has multiple meanings.

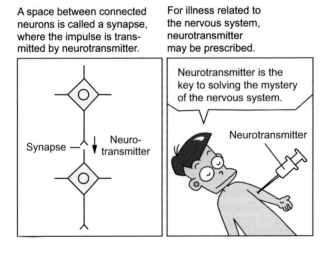

Fig. 2.3

When impulse arrives at the ending point of an axon, the axon releases a chemical substance called neurotransmitter (into the synapse), which carries impulse to the dendrite of the next neuron. The impulse transmission within a neuron is electrical, while that between neurons is chemical. In the case of the lower motor neuron, neurotransmitter carries impulse to the muscle (Figs. 2.6, 2.25).

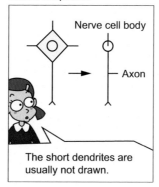

Fig. 2.4

The dendrites of multipolar neuron are very short compared with the axon; they are omitted in the schematic figure.

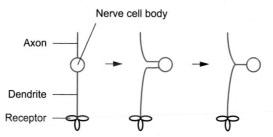

Fig. 2.5   Development of bipolar neuron to pseudounipolar neuron.

At the early stage of development, the sensory nerve's 1st neuron has an axon and a dendrite (bipolar neuron) (Fig. 3.1). In most cases, the dendrite and the axon are slightly fused to form a pseudounipolar neuron (Fig. 2.6). Then impulse is conveyed faster, bypassing the nerve cell body.

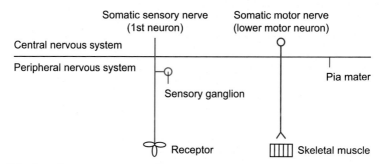

Fig. 2.6   Somatic sensory nerve, somatic motor nerve.

The somatic sensory nerve's 1st neuron [pseudounipolar neuron (Fig. 2.5)] receives impulse from the receptor around the skeletal muscle [skin, subcutaneous tissue, or skeletal muscle itself (Fig. 2.24)]. The 1st neuron's nerve cell body is called a sensory ganglion, because of the absolute rule that the nerve cell body located in the peripheral nervous system is a ganglion. Meanwhile, the somatic motor nerve's lower motor neuron [multipolar neuron (Fig. 2.4)] sends impulse to the skeletal muscle (voluntary muscle).

From now on, two pathways of the somatic sensory nerve and one pathway of the somatic motor nerve related with the spinal nerve will be explained.

The somatic sensory nerve

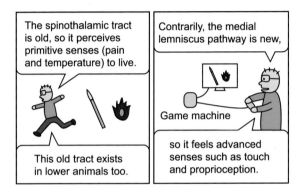

Fig. 2.7

The representative pathways of the somatic sensory nerve are the spinothalamic tract and medial lemniscus pathway. The spinothalamic tract conveys pain and temperature *(exactly, also crude touch)* (lower level sense). As a mnemonic, the SPinoTHalamic tract conveys pain caused by stabbing of a SPear and temperature measured by THermometer. The medial lemniscus pathway conveys touch *(exactly, discriminative touch)* and proprioception (higher level sense) (Fig. 3.29).

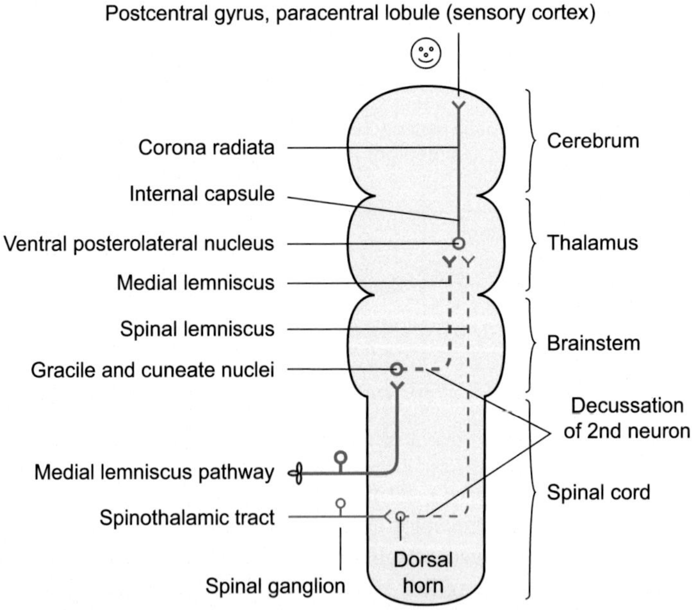

Postcentral gyrus, paracentral lobule (sensory cortex)

Corona radiata — Cerebrum

Internal capsule —

Ventral posterolateral nucleus — Thalamus

Medial lemniscus —

Spinal lemniscus —

Gracile and cuneate nuclei — Brainstem

Decussation of 2nd neuron

Medial lemniscus pathway

Spinothalamic tract — Spinal cord

Spinal ganglion — Dorsal horn

Fig. 2.8 Spinothalamic tract, medial lemniscus pathway.

The cerebrum, thalamus, and brainstem are drawn as three swellings. This figure is different from the three brain vesicles (forebrain, midbrain, hindbrain) during embryological development (Fig. 5.6).

The somatic sensory nerve is composed of three neurons. Starting point of the 1st neuron is the receptor that responds to the external stimulus to produce impulse (Fig. 2.6).

While the receptor of the spinothalamic tract is morphologically simple (free nerve ending), the receptor of the medial lemniscus pathway is complicated (encapsulated nerve ending). The reason is that the higher level the sense (Fig. 2.7), the bigger and more complicated the shape. By the same principle, the medial lemniscus pathway is thicker than the spinothalamic tract (Figs. 2.11, 2.12).

Regarding the two pathways together, the 1st neuron passes the spinal nerve (dorsal root) and forms the spinal ganglion, a kind of sensory ganglion (Figs. 2.6, 3.72)

The 1st neuron synapses with the 2nd neuron in the spinal cord (spinothalamic tract) or brainstem (medial lemniscus pathway). The higher level the sense, the higher the 2nd neuron's launching site.

In both pathways, the 2nd neuron decussates (crosses the median plane to the contralateral side) and ascends up to the ventral posterolateral nucleus of thalamus (Fig. 4.19). The authors emphasize the general rule of afferent nerves having three neurons. The key point of the rule is that the 2nd neuron decussates and ends at the thalamus (Tables 1, 2, 3).

Fig. 2.9

Specific neuroanatomy terms are used by professionals. For academic discussion, you should get used to the terms.

The ascending part of the 2nd neuron is called the lemniscus. The spinothalamic tract includes the spinal lemniscus. The spinal lemniscus from the SPINal cord to the THALAMus has determined the name, SPINoTHALAMic tract. The medial lemniscus pathway literally possesses the medial lemniscus (Fig. 2.8).

In general, the term "tract" is used in cases of combined origin and insertion like "spinothalamic tract," whereas the term "pathway" is used in other cases like "medial lemniscus pathway." Both tract and pathway consist of a bundle of neurons. Fig. 2.8 demonstrating the serial solitary neurons is not a portrayal of reality.

In Fig. 2.8, the 3rd neuron ascends along the internal capsule (Fig. 1.40) and corona radiata (Fig. 2.14). It ends at the postcentral gyrus (Fig. 1.26) and the continuous paracentral lobule (Fig. 1.28) (Table 1) *[exactly, their cerebral cortex (Fig. 1.31)]*. In other words, it arrives at the sensory cortex *(exactly, primary somatosensory cortex)* (Fig. 4.8).

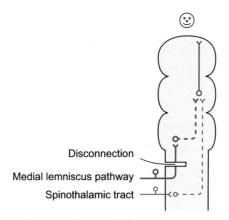

Fig. 2.10   Disconnection of spinal cord's right half.

Suppose that the right half of the spinal cord (upper cervical level) is disconnected (Fig. 1.65). In that case, the spinothalamic tract from the right (upper and lower) limbs is intact, while the medial lemniscus pathway from the right limbs is damaged (Fig. 2.14). As a result, the patient can feel pain and temperature from the right limbs, but cannot feel touch and proprioception from there. The reverse would be true, regarding the left limbs.

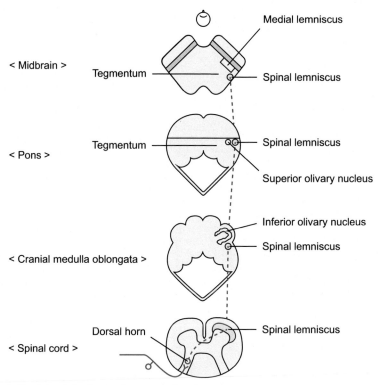

Fig. 2.11 Spinothalamic tract (transverse planes).

Let's explore the spinothalamic tract (Fig. 2.8) in the transverse planes of spinal cord and brainstem. Its 1st neuron synapses with the 2nd neuron at the dorsal horn that is a gray matter (Fig. 1.68).

The 2nd neuron decussates and ascends as the spinal lemniscus at the ventral and lateral funiculi that are white matter (spinal cord) (Fig. 1.68). *(Exactly, the spinal lemniscus is a brainstem structure including other pathways.)* The spinal lemniscus ascends dorsolateral to the inferior olivary nucleus (cranial medulla oblongata) (Fig. 1.58), lateral to the superior olivary nucleus (pons) (Fig. 1.54), and dorsolateral to the medial lemniscus (midbrain) (Fig. 2.12).

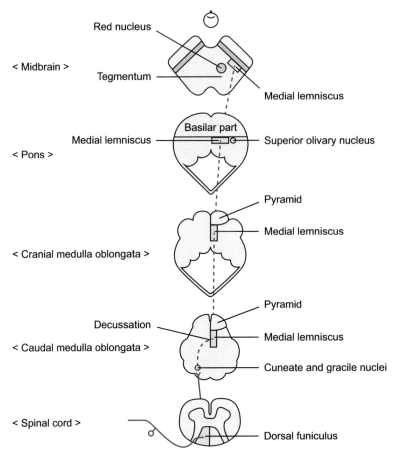

Fig. 2.12 Medial lemniscus pathway (transverse planes).

In the case of the medial lemniscus pathway (Fig. 2.8), the 1st neuron ascends in the dorsal funiculus (spinal cord) (Fig. 1.68). It synapses with the 2nd neuron at the cuneate and gracile nuclei (Fig. 2.14), which give rise to the cuneate and gracile tubercles (caudal medulla oblongata) (Figs. 1.58, 1.59, 2.13).

The 2nd neuron decussates at the same level, and ascends as the medial lemniscus that is dorsal to the pyramid (medulla oblongata) (Figs. 1.58, 1.59), dorsal to the basilar part (pons) (Fig. 1.54), and lateral to the red nucleus (midbrain) (Fig. 1.52).

Speaking repeatedly, the medial lemniscus is larger than the spinal lemniscus (Fig. 2.11). The higher level the sense (Fig. 2.7), the thicker the pathway (Fig. 2.8).

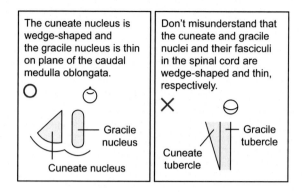

Fig. 2.13

The above cartoon shows the correct and incorrect etymologies of the cuneate and gracile nuclei. The "cuneus" in the occipital lobe is also wedge-shaped (Fig. 1.28) like the "cuneate" nucleus.

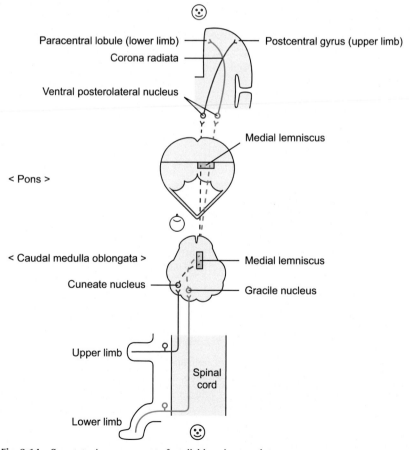

Fig. 2.14 Somatotopic arrangement of medial lemniscus pathway.

In terms of the somatotopic arrangement, compare the lower limb's sensory nerve (green line) with the upper limb's sensory nerve (purple line) of the medial lemniscus pathway. In the spinal cord, the lower limb is medial; in the caudal medulla oblongata, it is medial (gracile nucleus) too (Fig. 2.13).

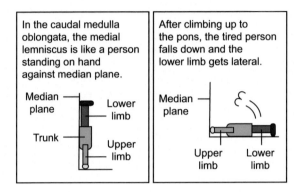

Fig. 2.15

After decussation in the caudal medulla oblongata, the lower limb becomes ventral; in the pons, it becomes lateral. It is a dramatic change of the somatotopic arrangement of the medial lemniscus pathway in the brainstem (Fig. 2.14).

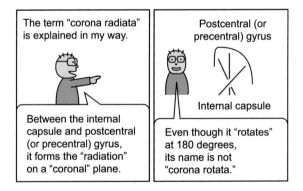

Fig. 2.16

In the ventral posterolateral nucleus and internal capsule, the lower limb remains lateral. But in the cerebral cortex, it becomes medial on the paracentral lobule (Fig. 1.28) because of the 3rd neurons' twisting in the corona radiata (Fig. 2.14). It is the other dramatic change, which results in the somatotopic arrangement in the postcentral gyrus and paracentral lobule (Fig. 4.8).

The somatic motor nerve

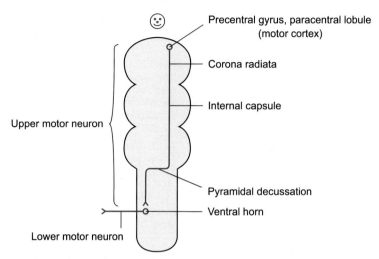

Fig. 2.17 Corticospinal tract.

The representative somatic motor nerve, corticospinal tract consists of the upper motor neuron and lower motor neuron (Fig. 2.6). The upper motor neuron originates from the precentral gyrus (Fig. 1.26) and the continuous paracentral lobule (Fig. 1.28), namely motor cortex *(exactly, primary motor cortex)* (Fig. 4.7). *(Exactly, the upper motor neuron may start from the other gyri of the frontal lobe and even from the postcentral gyrus.)*

The upper motor neuron descends by way of the corona radiata (Fig. 2.16) and internal capsule (Fig. 1.40). It decussates at the border between the caudal medulla oblongata and spinal cord. It is named "pyramidal" decussation in consideration of the "pyramid" of caudal medulla oblongata (Figs. 1.58, 2.19). The upper motor neuron descends further and meets the lower motor neuron at the ventral horn of spinal cord (Fig. 1.68).

[Exactly, this is the "lateral" corticospinal tract (majority), in which the upper motor neuron decussates and descends through the "lateral" funiculus (Fig. 2.19). In the "ventral" corticospinal tract (minority), the upper motor neuron does not decussate, keeps descending through the "ventral" funiculus (Fig. 1.68), and then decussates (or not) at the level of the lower motor neuron.]

In the corticospinal tract, the lower motor neuron passes through the ventral root of spinal nerve (Figs. 2.19, 3.72) and reaches the skeletal muscle (Fig. 2.6).

The upper and lower motor neurons are very long (almost 1 m for each). Think about the length between the cerebral cortex and the toe muscles. Those neurons must be the longest cells in body. A misunderstanding is that the upper and lower motor neurons correlate with the upper and lower limbs, respectively.

Fig. 2.18

Typically, the somatic sensory nerve consists of three neurons (Fig. 2.8); the somatic motor nerve consists of two neurons (Fig. 2.17). The sensory nerve synapses at the thalamus (Fig. 4.19) (Tables 1, 2, 3).

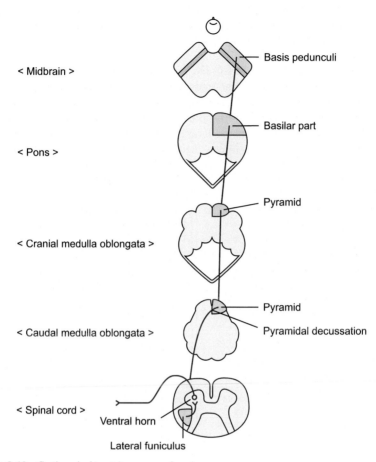

< Midbrain > — Basis pedunculi

< Pons > — Basilar part

< Cranial medulla oblongata > — Pyramid

< Caudal medulla oblongata > — Pyramid / Pyramidal decussation

< Spinal cord >
Ventral horn
Lateral funiculus

Fig. 2.19   Corticospinal tract (transverse planes).

The upper motor neuron of corticospinal tract (Fig. 2.17) descends through the basis pedunculi (midbrain), basilar part (pons), and pyramid (medulla oblongata). The basis pedunculi, basilar part, and pyramid are terms for both external and internal features (Figs. 1.52, 1.54, 1.58, 1.59). So is the pyramidal decussation which is visible at the ventral surface of the caudal medulla oblongata and spinal cord (Fig. 1.51).

In the ventral view, the large basilar part does not seem to contain the corticospinal tract unlike the basis pedunculi and pyramid (Fig. 1.51). This is because numerous axons from the pontine nuclei (Fig. 1.54) to the pontocerebellum (Fig. 4.37) hide the corticospinal tract.

Focusing on the midbrain, the spinothalamic tract and medial lemniscus pathway pass through the tegmentum (Figs. 2.11, 2.12); the corticospinal tract passes through the basis pedunculi. The tegmentum and basis pedunculi are collectively called the cerebral peduncle (Figs. 1.52, 1.57) that contains the sensory nerve to the cerebrum and the motor nerve from the cerebrum.

Since the upper motor neuron starts from the cerebral CORTex and ends at the destined level of SPINAL cord (Fig. 2.17), this somatic motor nerve is named CORTicoSPINAL tract.

In order to send impulse to CN IX, X, XII, the upper motor neuron starts from at the cerebral CORTex and ends at the medulla oblongata (Figs. 3.55, 3.61, 3.66). Because the medulla oblongata is called BULB, such somatic motor nerve is called CORTicoBULBar tract. Both the corticospinal and corticobulbar tracts pass the same structures (above the spinal cord) including the "pyramid" (Fig. 1.58). Therefore, the two tracts are called the "pyramidal" tract.

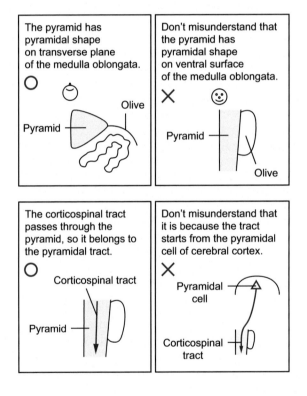

Fig. 2.20

The above cartoon shows the correct and wrong etymologies of the pyramid and pyramidal tract.

The reflex arc

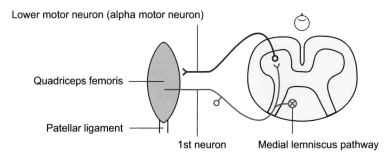

Lower motor neuron (alpha motor neuron)

Quadriceps femoris

Patellar ligament

1st neuron   Medial lemniscus pathway

Fig. 2.21   Reflex arc.

Look at the 1st neuron of the medial lemniscus pathway (Fig. 2.12) that conveys proprioception from a muscle. Few of the 1st neurons directly synapse with the lower motor neuron of the corticospinal tract (Fig. 2.19) in the spinal cord. It is called the "reflex arc" because it causes "reflex" (Fig. 2.22) and it looks like an "arc."

When the patellar ligament below the patella is tapped, the leg kicks forward by contraction of the quadriceps femoris.

This is known as knee jerk.

Fig. 2.22

The most evident sign of the reflex arc is the knee jerk, because the quadriceps femoris is a huge anterior thigh muscle and its innervating femoral nerve is thick (Fig. 3.81). Tapping the muscle tendon (patellar ligament) induces the muscle lengthening (Fig. 2.21). In order to prevent the excessive muscle lengthening, the reflex arc makes the muscle contract. The reflex arc is necessary for maintaining posture against external stimulus.

Because the reflex arc does not pass through the cerebral cortex (Fig. 4.1), the reflex happens autonomically. However, the reflex arc does not belong to the autonomic nerve; the reflex arc is not associated with the preganglionic neuron, postganglionic neuron, or smooth muscle (Fig. 2.25).

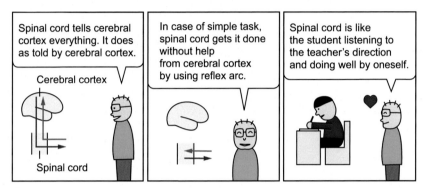

Fig. 2.23

The spinal cord as the center of reflex arc reminds us of self-studying student.

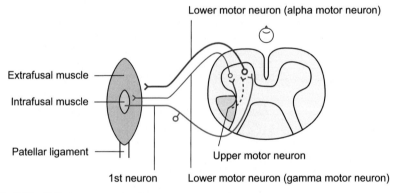

Fig. 2.24   Reflex arc, gamma motor neuron, upper motor neuron.

Skeletal muscle is composed of the extrafusal muscle (most part) and the intrafusal muscle (little part). The intrafusal muscle is identifiable only by microscope.

When the quadriceps femoris is lengthened, the intrafusal muscle makes impulse (proprioception) like a receptor and transmits the impulse to the spinal cord. In the spinal cord, the alpha motor neuron (most part of lower motor neuron) responds and sends the impulse to the extrafusal muscle making it contract (Figs. 2.21, 2.22).

Little part of the lower motor neuron is the gamma motor neuron which keeps the intrafusal muscle contracting, so the intrafusal muscle is not lengthened too easily.

If the lower motor neuron (both of the alpha and gamma motor neurons) is disconnected, the knee jerk will not happen.

If the upper motor neuron (Figs. 2.17, 2.19) is disconnected (Fig. 2.10), the gamma motor neuron will not get the impulse. Therefore, the intrafusal muscle will be lengthened too easily by tapping the patellar ligament; then the knee is exaggeratingly extended.

The reflex arc is intact regardless of the disconnected upper motor neuron (Fig. 2.21). You do not mind the regular upper motor neuron (dotted line in Fig. 2.24) that sends impulse to the alpha motor neuron (Figs. 2.17, 2.19).

In summary, disconnection of the lower motor neuron causes no knee jerk; disconnection of the upper motor neuron causes an exaggerated knee jerk.

The autonomic nerve

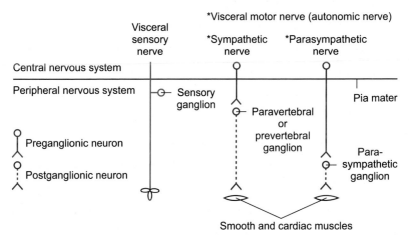

Fig. 2.25   Visceral sensory nerve, visceral motor nerve.

Like the somatic nerve, the visceral nerve resides both in the central and peripheral nervous systems. The visceral sensory nerve is similar to the somatic sensory nerve (Fig. 2.6). A difference is that the visceral sensory nerve delivers impulse from the receptor near the smooth and cardiac muscles instead of the skeletal muscle.

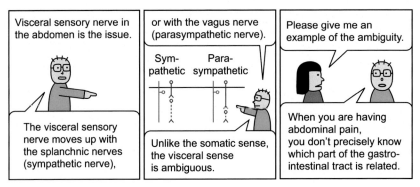

Fig. 2.26

Everyone comprehends the role of visceral sensory nerve (e.g., sensing hunger) by experience. The visceral sensory nerve accompanies the visceral motor nerve (sympathetic and parasympathetic nerves). The visceral sensory and visceral motor nerves are not discernible from each other during cadaver dissection (Fig. 2.25). Neither are the somatic sensory and somatic motor nerves (Fig. 2.6).

The visceral motor nerve innervates the smooth and cardiac muscles (involuntary muscle). The visceral motor nerve is referred to as "autonomic" nerve (Fig. 2.25), because it controls the muscle "autonomically" independent of one's will.

In the peripheral nervous system, the somatic motor nerve consists of a single neuron (Fig. 2.6), but the visceral motor nerve consists of two neurons: preganglionic and postganglionic neurons (fibers) (Fig. 2.25). Sometimes, the terms "neuron" and "fiber" are used interchangeably for the reason that a neuron's axon is long like a fiber.

The visceral motor nerve contains the ganglion that is the nerve cell body of the postganglionic neuron (dotted line). The postganglionic neuron of sympathetic nerve is longer than that of parasympathetic nerve. In other words, the sympathetic nerve has the ganglion close to the central nervous system, while the parasympathetic nerve has the ganglion close to the target muscle (Fig. 2.25).

Sympathetic nerve puts the body in a fight/flight mode, while parasympathetic nerve does the exact opposite.

Imagine that a student is caught smoking in the school restroom by a teacher, Mr. Simpson.

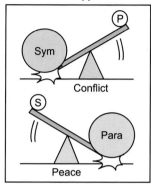

Noticing Mr. Simpson, the student will have an increased heart rate, trouble with digestion,

enlarged pupils, and decreased salivation.

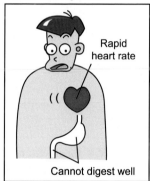

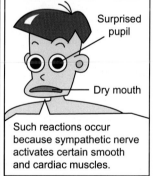

Fig. 2.27

The Sympathetic nerve is for Stimulated (war) state of the body (Figs. 3.21, 3.58), while the Parasympathetic nerve is for Peaceful state.

The sympathetic nerve is contained in T1−L2 (Fig. 2.31). (T1LTwo reminds us of a TILTed building in war state.) The parasympathetic nerve is contained in CN III (Fig. 3.18), CN VII (Fig. 3.37), CN IX (Fig. 3.55), CN X (Fig. 3.61), and S2−S4 (Fig. 2.35).

Overall, the four kinds of nerves (somatic sensory, somatic motor, visceral sensory, visceral motor nerves) (Figs. 2.6, 2.25, 3.68) make up a functional classification of the nervous system, while the central and peripheral nervous systems are its anatomical classification (Fig. 1.1).

The four kinds of nerves have synonyms (somatic afferent, somatic efferent, visceral afferent, visceral efferent nerves). The four are expanded by adding other subtypes (general and special). The resultant eight kinds *(exactly, seven kinds)* of nerves (general somatic afferent, special somatic afferent, ...) are not explained in this book.

Sympathetic nerve

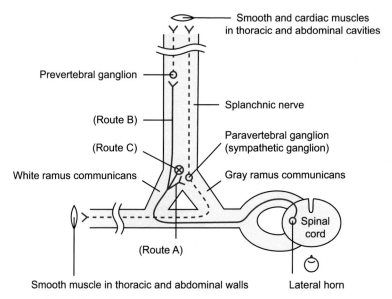

Fig. 2.28 Sympathetic nerve in spinal nerve.

At T1−L2 level, the preganglionic neuron of sympathetic nerve starts at the lateral horn of the spinal cord (Fig. 1.68). At S2−S4 level, the preganglionic neuron of parasympathetic nerve starts at the gray matter that corresponds to the upper level's lateral horn (Fig. 1.69).

In the above figure of the sympathetic nerve, the preganglionic neuron (solid line) and the postganglionic neuron (dotted line) seem similar in length. However, one must keep in mind that the postganglionic neuron is longer than the preganglionic neuron in the sympathetic nerve (Fig. 2.25). That is expressed with double wavy lines.

The preganglionic neuron passes through the ventral root and the trunk of spinal nerve, just like the somatic motor nerve (Figs. 2.19, 3.72). This neuron then travels through the white ramus communicans to reach a paravertebral ganglion (Fig. 2.29) where the neuron has three routes (A, B, C). The three routes are explained as follows.

(Route A: from T1−T4 to thoracic cavity) The preganglionic neuron synapses at the paravertebral ganglion. The postganglionic neuron then runs along the splanchnic nerve to innervate the smooth and cardiac muscles (lung, heart, ...) in the thoracic cavity (Fig. 2.31).

(Route A: from T1−L2 to thoracic, abdominal walls) After synapsing at the paravertebral ganglion, a small portion of the postganglionic neuron passes through the gray ramus communicans to rejoin the spinal nerve. The neuron accompanies the somatic motor nerve (Figs. 2.19, 3.72) to innervate the smooth muscle (blood vessel, sweat gland, hair) in the thoracic and abdominal walls (Figs. 3.73, 3.74).

(Route B: from T5−L2 to abdominal cavity) Without synapsing in the paravertebral ganglion, the preganglionic neuron runs along the splanchnic nerve and synapses at the prevertebral ganglion. The splanchnic nerve is situated between the paravertebral ganglion (Fig. 2.31) and prevertebral ganglion (Fig. 2.34). The postganglionic neuron then innervates the smooth muscle in the abdominal cavity.

In Fig. 2.28, the prevertebral ganglion is anterolateral to the vertebra containing the spinal cord (Fig. 1.8). Actually, the "prevertebral" ganglion is in "front" of the "vertebra." An example is the celiac ganglion, anterior to the abdominal aorta (Fig. 2.34).

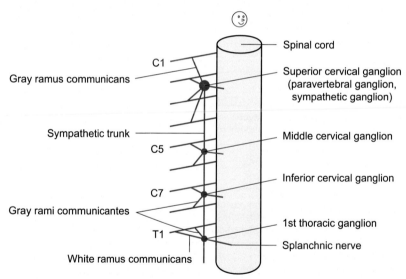

Fig. 2.29   Sympathetic trunk, adjacent structures.

(Route C: from the T1−L2 to head, neck, upper limb, pelvis, perineum, and lower limb) Without synapsing at the paravertebral ganglion, the preganglionic neuron takes an elevator known as the sympathetic trunk (Fig. 2.28) and synapses at the paravertebral ganglion of upper level (cervical ganglion) or lower level (lumbar or sacral ganglion). As an elevator, the "sympathetic" trunk connects the serial paravertebral ganglia vertically (Figs. 2.31, 2.35); that is why the paravertebral ganglia are also called the "sympathetic" ganglia.

The paravertebral ganglia have individual names according to their spinal nerve level, such as the 1st thoracic ganglion, connected to T1 (Fig. 2.31). At the cervical nerve level, the paravertebral ganglia fuse to form the inferior cervical ganglion (C7−C8), middle cervical ganglion (C5−C6), and superior cervical ganglion (C1−C4).

At the superior cervical ganglion, the elevated preganglionic neuron (symbol ⊗ in Fig. 2.28) synapses. [Symbol ⊗, borrowed from electromagnetism of physics, signifies a tail of an arrow going into the page (Figs. 1.3, 1.5, 2.21, 3.64).] The postganglionic neuron may proceed along the branches of the internal carotid artery (Fig. 1.3) and external carotid artery, and then it innervates smooth muscle in the head and neck.

A small portion of the postganglionic neuron (from the superior cervical ganglion) passes through the gray ramus communicans (Fig. 2.28) and participates in C1−C4. Then it innervates the smooth muscle in the tissue, where C1−C4 are distributed.

It is recommended that the readers compare Fig. 2.28 and Fig. 2.29 to confirm their understanding. For instance, T1 has both white and gray rami communicantes (plural form of ramus communicans), but C7 only has gray ramus communicans.

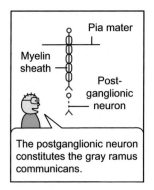

Fig. 2.30

Why are the rami communicantes white or gray (Fig. 2.28)? Usually, an axon is enclosed by the myelin sheath composed of white fat (Fig. 5.10). However, the postganglionic neuron is not enclosed; therefore, ramus communicans containing the postganglionic neuron is gray. This color difference is not recognizable during cadaver dissection (Fig. 2.29).

Autonomic nerve plexus

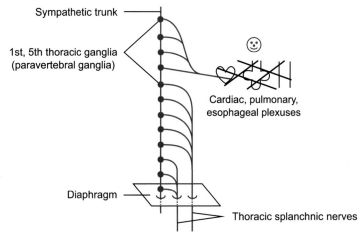

Fig. 2.31 Sympathetic nerve (thoracic cavity).

From the 1st–4th thoracic ganglia, the thoracic splanchnic nerves emerge heading for the cardiac and smooth muscles in the thoracic cavity (route A in Fig. 2.28). From the remaining 5th–12th thoracic ganglia, the thoracic splanchnic nerves pierce the diaphragm (Fig. 3.77) to approach the smooth muscle in the abdominal cavity (route B in Fig. 2.28) (Fig. 2.34).

The sympathetic nerve is mixed up with the parasympathetic nerve to form plexuses near the target organs (Figs. 2.33, 2.34, 3.57).

Fig. 2.32

Related to the parasympathetic nerve, CN III, VII, IX send parasympathetic impulse to the head and neck (Fig. 1.64). On the other hand, CN X (vagus nerve) sends parasympathetic impulse to the thoracic and abdominal cavities (Figs. 2.33, 2.34, 3.53, 3.57).

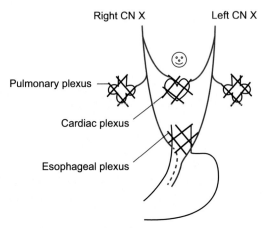

Fig. 2.33 Parasympathetic nerve (thoracic cavity).

A branch of CN X in the neck joins the cardiac plexus; another branch of CN X in the thoracic cavity joins the pulmonary plexus. The main trunk of CN X joins the esophageal plexus (Fig. 3.57) and enters the abdominal cavity to join the celiac plexus, and so on (Fig. 2.34).

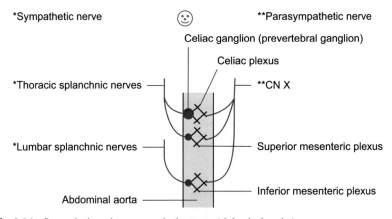

Fig. 2.34 Sympathetic and parasympathetic nerves (abdominal cavity).

In the abdominal cavity, the lumbar splanchnic nerves from the lumbar ganglia, as well as the thoracic splanchnic nerves from the thoracic ganglia (sympathetic nerve) (route B in Fig. 2.28) (Fig. 2.31) and CN X (parasympathetic nerve) (Fig. 2.33) form the celiac, superior mesenteric, and inferior mesenteric plexuses.

In the three plexuses, there exist the bilateral prevertebral ganglia that are nerve cell bodies of the postganglionic neurons of sympathetic nerve (Fig. 2.28). The celiac ganglion is the largest prevertebral ganglion, and the superior cervical ganglion is the largest paravertebral ganglion (Fig. 2.29).

From the celiac, superior mesenteric, and inferior mesenteric plexuses, the sympathetic and parasympathetic nerves travel along with the branches of their respective arteries (celiac trunk, superior mesenteric artery, and inferior mesenteric artery). Eventually, the sympathetic and parasympathetic impulses are appropriately delivered to the abdominal organs.

In the case of the gastrointestinal tract, the parasympathetic ganglion and smooth muscle coexist in its wall, because the postganglionic neuron of parasympathetic nerve is extremely short (Fig. 2.25). *(Exactly, the postganglionic neuron of sympathetic nerve may synapse with the additional 3rd neuron in the gastrointestinal tract.)*

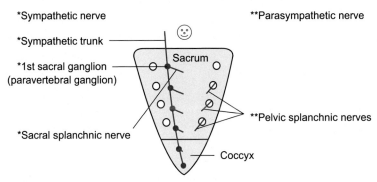

Fig. 2.35 Sympathetic and parasympathetic nerves (pelvic cavity, perineum).

In the pelvic cavity and perineum, the sacral splanchnic nerves from the sacral ganglia (sympathetic nerve) (Route C in Fig. 2.28) and the pelvic splanchnic nerves from S2−S4 (parasympathetic nerve) form plexuses and are responsible for the smooth muscle. Except the pelvic splanchnic nerve, the splanchnic nerves convey sympathetic impulse (Figs. 2.28, 2.29) and are roughly named after their corresponding paravertebral ganglia (Figs. 2.31, 2.34).

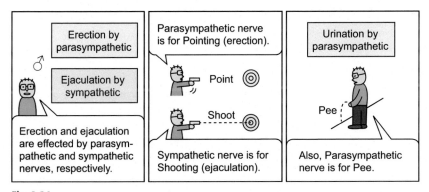

Fig. 2.36

In the pelvic cavity and perineum, the main target of sympathetic and parasympathetic nerves is the smooth muscle for erection, ejaculation of male and the smooth muscle for urination of both sexes.

All sympathetic and parasympathetic nerves are influenced by the hypothalamus (Fig. 4.26), the headquarters of autonomic nerve. The neuron from the hypothalamus may reach the autonomic nuclei in the brainstem and spinal cord, through relay of the reticular formation (Fig. 4.28).

Chapter 3

The cranial nerve, the spinal nerve

The cranial nerves consists of 12 pairs of nerves, most of which emerge from the brainstem. In terms of function, the cranial nerve contains somatic sensory nerve, somatic motor nerve, visceral sensory nerve, and visceral motor nerve. Correspondingly, each cranial nerve contains the nucleus (or nuclei) in the central nervous system, and may contain the ganglion (or ganglia) in the peripheral nervous system. The functions and locations of the nuclei and ganglia are discussed in detail. Readers can be familiarized with the nuclei with assistance from the stained slices of the brainstem. In succession, the spinal nerves (cervical, thoracic, lumbar, and sacral nerves) from the spinal cord are studied, with respect to the components of the spinal nerves (somatic sensory nerve and somatic motor nerve). For full understanding of the cranial and spinal nerves, regional anatomy knowledge is necessary.

The cranial nerve

Cranial nerve I

The first two cranial nerves, CN I and II, are often regarded as the extended parts of the brain. Technically speaking, CN I and II are the extensions of cerebrum and thalamus, respectively (Fig. 1.61). CN I and II belong to the central nervous system, because the two nerves are enclosed by the pia mater which covers the central nervous system (Figs. 1.7, 3.83). This explains why CN I and II have no sensory ganglion, a structure of the peripheral nervous system (Fig. 2.6) (Table 2).

CN I is discussed briefly in this neuroanatomy book, since it is rather close to the neurophysiology field.

Visually Memorable Neuroanatomy for Beginners. DOI: https://doi.org/10.1016/B978-0-12-819901-5.00003-X

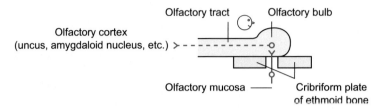

Fig. 3.1 Olfactory pathway.

The 1st neuron of CN I originates from receptor of the olfactory mucosa in the upper nasal cavity. The 1st neuron [short bipolar neuron (Fig. 2.5)] synapses with the 2nd neuron in the olfactory bulb. The 2nd neuron runs through the olfactory tract. The olfactory bulb and tract are beneath the frontal lobe (Fig. 1.5), so they are located on the anterior cranial fossa (Fig. 1.25).

Since CN I is the extension of the cerebrum, the 2nd neuron does not go to the thalamus (Fig. 3.83), but directly to the olfactory cortex. That is quite different from the general rule of afferent nerves (Fig. 2.8). (The readers may ignore CN I in Table 2.) Since the 2nd neuron does not decussate, it goes to the ipsilateral olfactory cortex, which is another difference. *[Exactly, small part of the 2nd neurons in the bilateral olfactory bulbs communicate through the anterior commissure (Fig. 1.44), which is not a typical decussation of afferent nerve.]*

The olfactory cortex is scattered in the temporal lobe (Fig. 1.5), so it is rare to encounter a patient with olfactory malfunction caused by the localized brain damage. Examples of the olfactory cortex are the uncus (Fig. 1.28) and its inside structure, amygdaloid nucleus (Fig. 1.39) which belongs to the limbic system (Fig. 4.14). Smell is the only sense that directly gets to the limbic system. Considering the limbic system's function, smell such as perfume must strongly affect memory and emotion.

Cranial nerve II

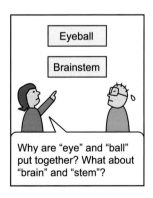

Fig. 3.2

The authors do not know why the eyeball and brainstem are spelled without spacing. CN II comes from the eyeball, without passing the brainstem (Fig. 3.5).

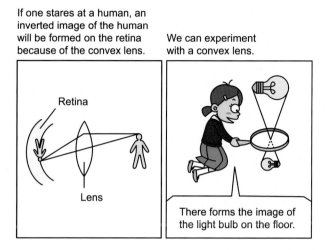

Fig. 3.3

Light entering the eyeball is refracted by the lens. As a result, the inverted image is projected onto the retina.

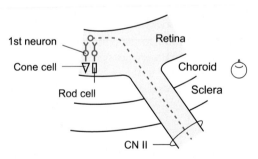

Fig. 3.4 Retina, adjacent structures.

The image stimulates the cone cell and rod cell of the retina. The authors regard the two kinds of cells as receptors, not as neurons. The receptors are located posterior to the 1st neurons, so that image should penetrate the 1st neurons to reach the receptors. This is such an unusual architecture.

The 1st neuron is bipolar and very short. It synapses with the 2nd neuron, which passes the anterior part of retina and exits the eyeball to become CN II.

In the olfactory and visual pathways, the 1st neuron (bipolar neuron) meets the 2nd neuron in the olfactory bulb (Fig. 3.1) and retina. So the olfactory bulb and retina appear homologous.

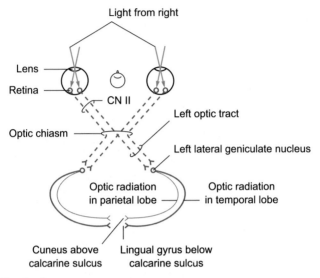

Fig. 3.5 Visual pathway.

In the conventional sensory nerve, the 2nd neuron decussates (Fig. 2.8). But in the visual pathway, only half of the 2nd neurons decussate at the optic chiasm (Fig. 1.62). Consequently, light entering from the right is sent to the left optic tract and the left lateral geniculate nucleus of thalamus (Fig. 4.19) (Table 2). As noted, CN II is the extension of thalamus (Figs. 1.61, 3.83).

The 2nd neuron of CN I travels through the olfactory tract (Fig. 3.1); the 2nd neuron of CN II travels through the optic tract. The term "tract" refers to a bundle of axons in the central nervous system. Therefore, the olfactory and optic tracts that appear to belong to the peripheral nervous system (Figs. 1.5, 1.62) are in fact part of the central nervous system.

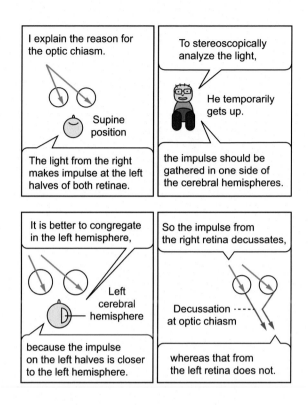

Fig. 3.6

The light from the right reaches the left visual cortex above and below the calcarine sulcus (Figs. 1.28, 3.5, 3.10); as a result, it is recognized how far the light comes from.

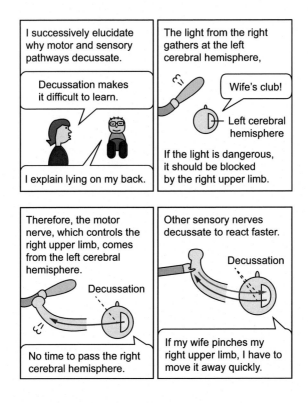

Fig. 3.7

The visual pathway is likely to cause the decussation of the motor nerves (Fig. 2.17) and other sensory nerves (Fig. 2.8).

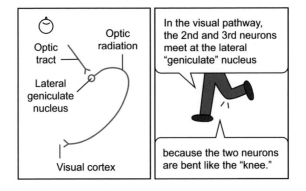

Fig. 3.8

The above cartoon depicts etymology of the lateral geniculate (meaning knee) nucleus (Fig. 3.5).

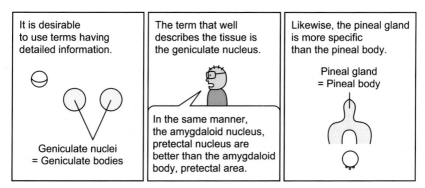

Fig. 3.9

The authors prefer the term "geniculate nucleus" to "geniculate body." The geniculate nucleus employs more concrete and more consistent term "nucleus" as a thalamic nucleus (Fig. 1.43).

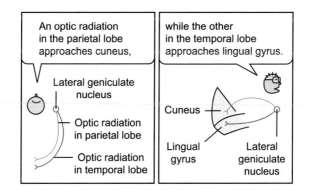

Fig. 3.10

The 3rd neurons carry impulse to the visual cortex in the occipital lobe (Figs. 1.28, 4.11). The visual cortex is much broader than the lateral geniculate nucleus, so the 3rd neurons spread like a hand fan and are thus called optic "radiation" (Fig. 3.5). A similar situation happens in the corona "radiata" that spreads to the broad cerebral cortex (Fig. 2.14).

In the above cartoon and other figures of this book, a nerve cell body gives off two axons. No such neuron exists in reality (Fig. 2.2). The authors have excluded another nerve cell body in order to create simpler figures.

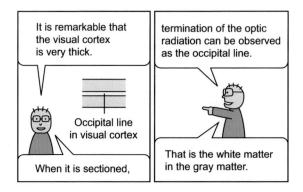

Fig. 3.11

The occipital line (line of Gennari) is the white matter running through the visual cortex. The occipital line is white due to abundant myelinated axons (Fig. 5.10) of the optic radiation (Fig. 3.10).

Cranial nerve III

The sensory and motor neurons of CN III−XII exist both in the central and peripheral nervous systems (Figs. 2.6, 2.25, 3.68).

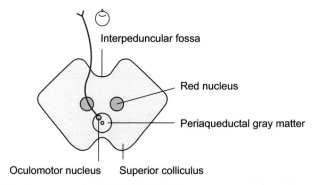

Fig. 3.12   Somatic motor nerve of CN III (midbrain).

The oculomotor nucleus resides in the periaqueductal gray matter at the level of the superior colliculus. CN III emerges from the interpeduncular fossa (Figs. 1.52, 1.62).

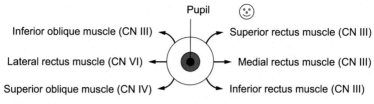

Fig. 3.13 Action of extraocular muscles (innervation).

CN III innervates the four extraocular muscles (superior rectus, medial rectus, inferior rectus, and inferior oblique muscles) (Figs. 3.19, 3.22).

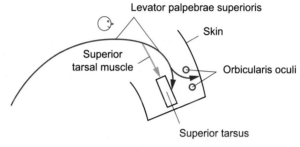

Fig. 3.14 Levator palpebrae superioris, superior tarsal muscle in superior eyelid.

CN III also controls the levator palpebrae superioris (muscle to elevate the superior eyelid) (Figs. 3.19, 3.39).

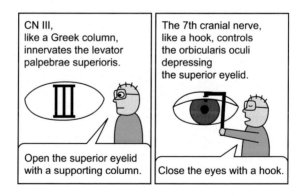

Fig. 3.15

CN III opens the superior eyelid; CN VII closes the superior and inferior eyelids. This distinction is because the orbicularis oculi is one of the facial muscles (Figs. 3.14, 3.35).

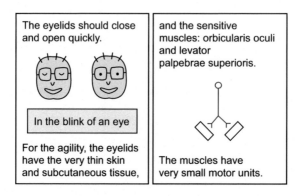

Fig. 3.16

A motor unit is defined as a lower motor neuron and its innervating muscle cells (Fig. 2.6). The fewer muscle cells there are in a motor unit, the more finely the muscle contracts. The delicate eyelid muscles (Fig. 3.14) and extraocular muscles (Fig. 3.13) have small motor unit.

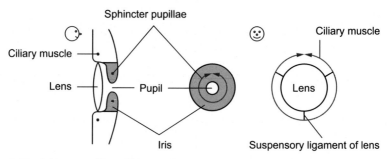

Fig. 3.17 Sphincter pupillae, ciliary muscle.

In addition, CN III contains the parasympathetic nerve, which constricts the pupil (sphincter pupillae) and thickens the lens (ciliary muscle) (Fig. 3.19). The light reflex which causes pupil constriction is explained in the below figure.

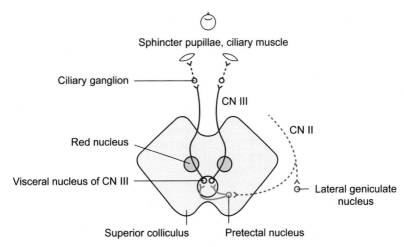

Fig. 3.18   Light reflex.

We know that the 2nd neuron of the visual pathway goes to the lateral geniculate nucleus (Fig. 3.5). During the light reflex, a small portion of the 2nd neurons go to the preTECTal nucleus, which is ventral (anterior) to the superior colliculus. Remember that the TECTum of midbrain includes the superior colliculus (Fig. 1.52).

The next neuron from the pretectal nucleus reaches the ipsilateral or contralateral visceral nucleus of CN III (Edinger-Westphal nucleus), which is in contact with the oculomotor nucleus (Fig. 3.12).

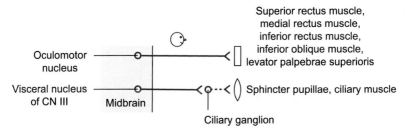

Fig. 3.19   CN III.

The preganglionic neuron from the visceral nucleus of CN III synapses with the postganglionic neuron at the ciliary ganglion (a parasympathetic ganglion) (Fig. 2.25). The postganglionic neuron then innervates the sphincter pupillae to yield the light reflex. The postganglionic neuron also innervates the "ciliary" muscle (Figs. 3.17, 3.18); hence, the term "ciliary" ganglion.

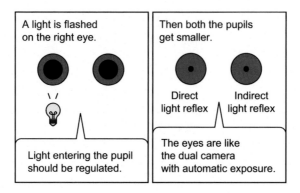

Fig. 3.20

When light is given to the right pupil, the bilateral pupils are constricted. Ipsilateral pupil constriction is triggered by direct light, while contralateral pupil constriction by indirect light. The direct and indirect light reflexes simultaneously occur because of two reasons: (1) the 2nd neuron from the retina decussates or does not in the optic chiasm (Fig. 3.5) and (2) the neuron from the pretectal nucleus decussates or does not (Fig. 3.18).

The sympathetic nerve also affects the eye. In addition to the levator palpebrae superioris, the superior tarsal muscle contributes to the elevation of the superior eyelid (Fig. 3.14). While the levator palpebrae superioris is a skeletal muscle, the Superior tarsal muscle is a Smooth muscle innervated by the Sympathetic nerve (Fig. 3.39) from the Superior cervical ganglion (Fig. 2.29).

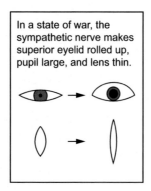

Fig. 3.21

The sympathetic nerve affects the eyes in a state of war (Fig. 2.27), opposite to the parasympathetic nerve of CN III (Fig. 3.18). The sympathetic nerve allows a soldier to see enemies in a wide range (with the elevated superior eyelid) (Fig. 3.14), enemies in a dark field (with the large pupil), and enemies in a far distance (with the thin lens) (Fig. 3.17).

Cranial nerve IV

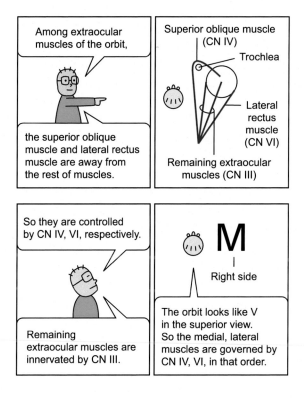

Fig. 3.22

CN IV and VI control the other two extraocular muscles, the superior oblique muscle and lateral rectus muscle, respectively. In the 2nd frame of the above cartoon, a "trochlea" (meaning pulley) is for the superior oblique muscle; the innervating nerve is "trochlear" nerve (CN IV). The lateral rectus muscle induces "abduction" of the pupil (Fig. 3.13); the innervating nerve is "abducens" nerve (CN VI).

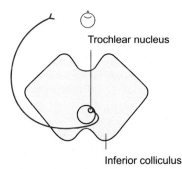

Fig. 3.23   CN IV (midbrain).

The trochlear nucleus is located in the periaqueductal gray matter at the level of the inferior colliculus, where the red nucleus is absent. The trochlear nucleus is directly inferior to the oculomotor nucleus which is at the level of the superior colliculus (Fig. 3.12).

Unlike other lower motor nerves (Fig. 2.17), the lower motor neuron of CN IV decussates in the central nervous system. This decussation is related to the vestibuloocular reflex (Fig. 3.50).

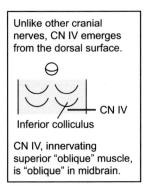

Unlike other cranial nerves, CN IV emerges from the dorsal surface.

Inferior colliculus

CN IV

CN IV, innervating superior "oblique" muscle, is "oblique" in midbrain.

Fig. 3.24

In general, cranial nerves emerge from the ventral surface of brainstem, but CN IV emerges from the dorsal surface (Fig. 1.62). The emerging area of CN IV is inferior to the inferior colliculus, while the trochlear nucleus is located at the level of the inferior colliculus. This implies that the lower motor neuron of CN IV is slightly oblique inside the midbrain (Fig. 3.23).

Cranial nerve VI

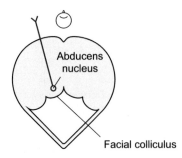

Abducens nucleus

Facial colliculus

Fig. 3.25   CN VI (pons).

CN VI related to eyeball movement is explained prior to CN V. CN VI emerges from the border between the pons and medulla oblongata (in detail, pyramid) (Fig. 1.62), while the abducens nucleus is located in the facial colliculus of the pons (Fig. 1.54). This implies that the lower motor neuron of CN VI is oblique inside the pons like that of CN IV (Fig. 3.24).

The oculomotor nucleus (Fig. 3.12), trochlear nucleus (Fig. 3.23), abducens nucleus (official terms: nuclei of oculomotor nerve, trochlear nerve, abducens nerve) are affected by the vestibular nucleus (Figs. 3.48, 3.50) and by the superior colliculus (Fig. 4.45).

Cranial nerve V

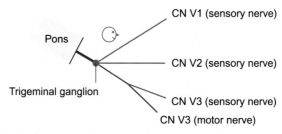

Fig. 3.26   CN V1, V2, V3.

CN V emerges from the basilar part of the pons (Fig. 1.62), and divides into CN V1, V2, V3 just after the trigeminal ganglion. Among the three components, only CN V3 contains the motor nerve (Fig. 3.30).

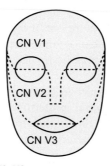

Fig. 3.27   Skin areas of CN V1, V2, V3.

The three skin areas innervated by the ophthalmic nerve (CN V1) (main branch, "frontal" nerve), "maxillary" nerve (CN V2), and "mandibular" nerve (CN V3) are on the "frontal" bone, "maxilla," and "mandible," respectively. CN V2 and V3 also relay impulses (e.g., terrible toothache) from "maxillary" teeth and "mandibular" teeth (Fig. 3.30).

A branch of CN V3 is the lingual nerve, which receives general sense from the tongue (anterior 2/3) (Fig. 3.30). An example of general sense is pain felt when the tongue is bitten, rather than taste which is a special sense (Fig. 3.33).

The sensory nerve of CN V contains the trigeminal ganglion (common sensory ganglion of CN V1, V2, V3) (Figs. 2.6, 3.26, 3.28), much like the sensory nerve of the spinal nerve containing the spinal ganglion (Fig. 3.72).

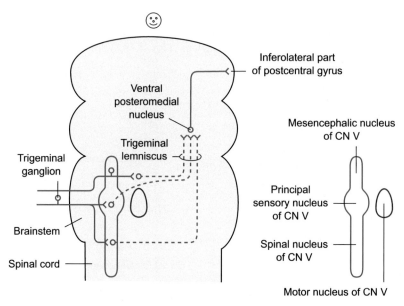

Fig. 3.28 Sensory nerve of CN V (trigeminothalamic tract).

The trigeminal ganglion belongs to the 1st neuron carrying pain, temperature, and touch (Fig. 3.26). The 1st neuron synapses with the 2nd neuron either at the "spinal" nucleus of CN V in the "spinal" cord and medulla oblongata (relaying pain and temperature), or at the principal sensory nucleus of CN V in the pons (relaying touch). A similar situation occurs in the spinothalamic tract (relaying pain and temperature) and the medial lemniscus pathway (relaying touch) (Fig. 2.8).

The 2nd neuron from the nuclei decussates and ascends as the "trigeminal" lemniscus to end at the ventral posteromedial nucleus of "thalamus" (Fig. 4.19). Therefore, the pathway name is "trigeminothalamic" tract (Table 2). In the brainstem, the trigeminal lemniscus accompanies the medial lemniscus (Fig. 2.12).

Fig. 3.29

The trigeminothalamic tract is also responsible for the proprioception of masticatory muscles. The receptor of proprioception is located in the muscle (Fig. 2.24), tendon, and joint. The 1st neuron passes CN V3 in the peripheral nervous system (Fig. 3.30).

The 1st neuron carrying the proprioception of masticatory muscles is exceptional, since it has a nucleus (mesencephalic nucleus of CN V) instead of a ganglion (Fig. 3.30). Higher level sense like proprioception tends to be handled by higher level structure of the nervous system (Fig. 2.8). The 1st neuron immediately synapses with the 2nd neuron, which decussates and ascends as another member of the trigeminal lemniscus (Fig. 3.28).

The 3rd neurons from the ventral posteromedial nucleus, which is medial to the ventral posterolateral nucleus (Fig. 4.19), are twisted at 180 degrees in the corona radiata (Figs. 2.14, 2.16). Consequently, the sensory nerve innervating the face (Fig. 3.27) occupies the inferolateral part of the postcentral gyrus (Fig. 3.28). This part is large, thus resulting in the large face of sensory homunculus (Fig. 4.8).

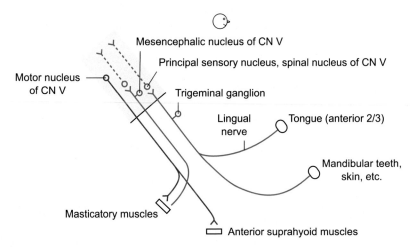

Fig. 3.30   CN V3.

The lower "motor" neuron from the "motor" nucleus of CN V in the pons (Figs. 3.28, 3.32) innervates the masticatory muscles and anterior suprahyoid muscles (Fig. 3.36). The suprahyoid muscles are above the hyoid bone (Fig. 3.67) which is palpable over the larynx.

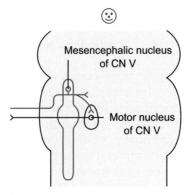

Fig. 3.31   Reflex of CN V3.

Some of the 1st neurons carrying the proprioception of the masticatory muscles go directly to the motor nucleus of CN V. As mentioned just before, the nucleus is responsible for the masticatory muscles (Fig. 3.30). As a type of reflex arc (Fig. 2.21), this circuit is needed for controlling bite strength of the masticatory muscles according to the food's firmness (Fig. 3.29).

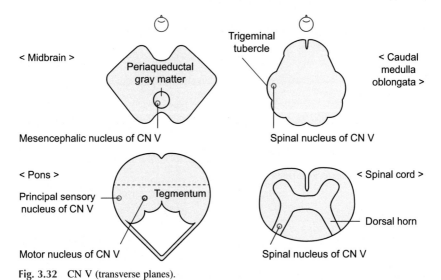

Fig. 3.32   CN V (transverse planes).

The above transverse planes demonstrate locations of the nuclei of CN V. The mesencephalic nucleus of CN V is located in the periaqueductal gray matter (midbrain); the principal sensory nucleus and motor nucleus of CN V are located in the tegmentum (pons) (Fig. 3.28).

The spinal nucleus of CN V extends to the dorsal horn of spinal cord and ends at C3 level; therefore, it can be found in the medulla oblongata (Fig. 3.28). The spinal nucleus of "trigeminal" nerve is inside the "trigeminal" tubercle of the medulla oblongata (Figs. 1.58, 3.54, 3.59).

Pathway of the spinal nucleus of CN V (synapsing at the dorsal horn) (Fig. 3.28) relays pain and temperature. So does the spinothalamic tract (also synapsing at the dorsal horn) (Fig. 2.11).

CN V in the central nervous system comprises three sensory nuclei and one motor nucleus (Fig. 3.28). The multiple and long nuclei are connected with the notably thick CN V in the peripheral nervous system (Figs. 1.62, 3.26).

Cranial nerve VII

CN VII in the cranial cavity (Fig. 1.63) enters the temporal bone and intricately divides within it (Fig. 3.35).

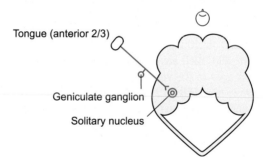

Fig. 3.33 Sensory nerve of CN VII (cranial medulla oblongata).

The solitary nucleus of CN VII receives special sense, taste from the tongue (anterior 2/3) (Fig. 3.35). (Taste bud in the tongue is receptor for this special sense.) Fun reason for this function is that the solitary nucleus surrounding the central white matter looks like a tasty doughnut; the "solitary" nucleus makes the surrounded white matter "solitary."

Actually, most neurons of CN VII are bent around the "geniculate" ganglion (Fig. 3.35) unlike the above figure. Similar bending happens around the lateral "geniculate" nucleus (Fig. 3.8).

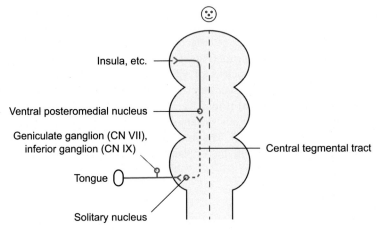

Fig. 3.34 Taste pathway.

In the taste pathway, the solitary nucleus is the start of the 2nd neuron that ascends as the "central tegmental" tract that is literally located at the "center of tegmentum" (Figs. 1.52, 1.54). The 2nd neuron does not decussate; in the olfactory pathway, the 2nd neuron does not decussate either (Fig. 3.1) (Table 2). This is because taste and smell are closely related to each other for eating.

The 2nd neuron goes to the ventral posteromedial nucleus of thalamus, like the trigeminothalamic tract (Fig. 3.28). The 3rd neuron subsequently goes to the insula (Fig. 1.27) and the opercular part of the frontal lobe (Fig. 1.26) (Table 2). The "solitary" nucleus is connected with the "solitary" insula. The insula and the opercular part are not at a distance from the sensory cortex of the head including the tongue (Fig. 4.8).

In this book, the 2nd and 3rd neurons of the cranial nerve (sensory nerve) are depicted in the cases of CN II (Fig. 3.5), CN V (Fig. 3.28), CN VIII (Fig. 3.51) as well as CN VII, CN IX (taste pathway), because they are dissimilar to the ordinary 2nd and 3rd neurons of the spinal nerve (Fig. 2.8).

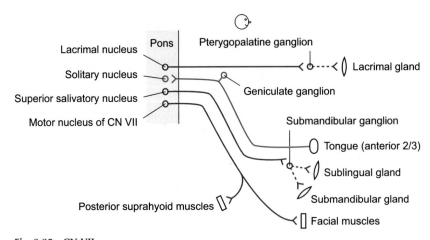

Fig. 3.35 CN VII.

The lower "motor" neuron of the "motor" nucleus of CN VII innervates the facial muscles and posterior suprahyoid muscles. The "facial" muscles are the nomenclatural origin of the innervating CN VII, "facial" nerve. The "motor" nucleus of CN VII reminds us of the "motor" nucleus of CN V, which innervates the masticatory muscles and anterior suprahyoid muscles (Fig. 3.30). The suprahyoid muscles contract to elevate the larynx (Fig. 3.67); you can touch the larynx elevating during swallowing.

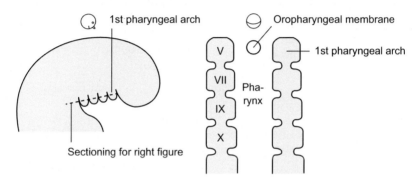

Fig. 3.36　Pharyngeal arches.

The complicated muscles, innervated by CN V, VII, IX, X, can be categorized by use of the pharyngeal arches.

The pharyngeal arches are formed during the developmental stage of the head and neck. As their names indicate, inside of the pharyngeal arches is the pharynx. The oropharyngeal membrane ruptures to become the fauces between the oral cavity and pharynx. The skeletal muscles in the 1st, 2nd, 3rd, 4th pharyngeal arches are innervated by CN V, VII, IX, X, respectively.

Even after birth, the masticatory muscles and anterior suprahyoid muscles from the 1st pharyngeal arch are innervated by CN V (specifically, CN V3) (Fig. 3.30); the facial muscles and posterior suprahyoid muscles from the 2nd pharyngeal arch are innervated by CN VII (Fig. 3.35). *[Exactly, the additional muscles, innervated by Trigeminal nerve (CN V) and Seventh cranial nerve (CN VII), are Tensor muscles (tensor tympani, tensor veli palatini) and Stapedius, respectively.]*

In succession, a muscle in pharynx from the 3rd pharyngeal arch is innervated by CN IX (Fig. 3.55); the muscles in palate, pharynx, and larynx from the 4th pharyngeal arch are innervated by CN X (Fig. 3.61).

CN V, VII, IX, X of the pharyngeal arches should not be confused with CN III, VII, IX, X of the parasympathetic nerve (Fig. 3.68).

CN V, VII, IX, X include Both the sensory nerve and motor nerve. (So they are hot potatoes.) On the other hand, CN I, II, VIII include only the Sensory nerve; CN III, IV, VI, XI, XII include only the Motor nerve (Figs. 1.62, 2.6, 2.25, 3.68). It can be memorized with the following sentence with 12 words: "Small Ships Make Money, But My Brother Says Big Boats Make More."

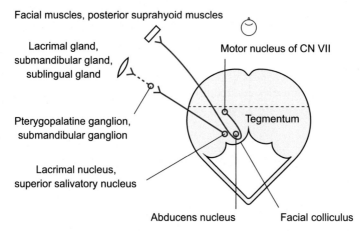

Facial muscles, posterior suprahyoid muscles

Lacrimal gland,
submandibular gland,
sublingual gland

Motor nucleus of CN VII

Pterygopalatine ganglion,
submandibular ganglion

Tegmentum

Lacrimal nucleus,
superior salivatory nucleus

Abducens nucleus   Facial colliculus

Fig. 3.37   Motor nerve of CN VII (pons).

The motor nucleus of CN VII is situated in the tegmentum of pons. Its lower motor neuron travels dorsally and loops around the abducens nucleus (Fig. 3.25). Finally, it travels ventrally and a bit caudally to exit between the pons and medulla oblongata (Fig. 1.62).

The lower motor neuron from the motor nucleus of "facial" nerve determines the name "facial" colliculus (Fig. 1.54), which contains the abducens nucleus (Fig. 3.25). When the colliculus is named, the superficial nerve is regarded rather than the deep one.

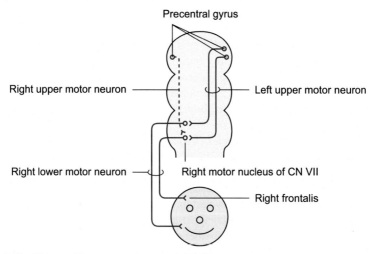

Precentral gyrus

Right upper motor neuron

Left upper motor neuron

Right lower motor neuron

Right motor nucleus of CN VII

Right frontalis

Fig. 3.38   Upper and lower motor neurons of CN VII.

The left upper motor neuron of CN VII arises from the precentral gyrus (inferolateral part) (Fig. 1.26) and decussates to encounter the right motor nucleus of CN VII (Fig. 3.37). But in the case of the right frontalis (facial muscle in forehead), the upper motor neuron decussates or does not (dotted line). As a consequence, the right frontalis is not paralyzed even if the left upper motor neuron is disconnected (Fig. 2.10).

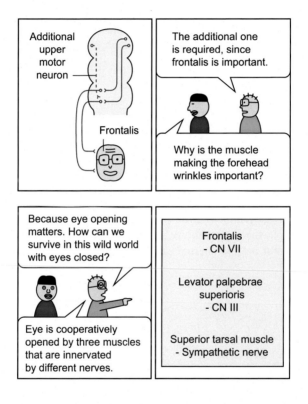

Fig. 3.39

For the upper motor neuron of CN VII to not decussate (Fig. 3.38) is likely to occur due to the significant function of the frontalis that assists eye opening. Review the other muscles opening eyes and their nerves (Figs. 3.14, 3.15, 3.21).

Excluding CN VII and XII, the upper motor neuron of the cranial nerve frequently does not decussate. This phenomenon is dissimilar to the upper motor neuron of the spinal nerve (Fig. 2.17).

The preganglionic neuron from the lacrimal nucleus synapses with the pterygopalatine ganglion to innervate the lacrimal gland. (The term "pterygopalatine" came from the bony structure, "pterygopalatine" fossa formed by "pterygoid" process and "palatine" bone, where the ganglion resides.) Additionally, the preganglionic neuron from the superior salivatory nucleus synapses with the submandibular ganglion to innervate the submandibular and sublingual glands (Fig. 3.35).

The lacrimal nucleus and superior salivatory nucleus are not distinguishable in the transverse plane of the pons (Fig. 3.37). Functionally, they are the parasympathetic components of CN VII, causing secretion of tears and saliva in the movie theater and restaurant (peaceful places for a date) (Fig. 2.27).

Cranial nerve VIII

CN VIII (vestibulocochlear nerve) involves the vestibular nerve (small portion) and cochlear nerve (large portion), which transmit balance sense and sound from the internal ear, separately.

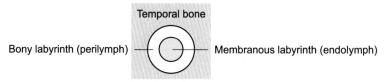

Fig. 3.40   Bony and membranous labyrinths.

If the temporal bone is soil, the bony labyrinth is a tunnel in the soil, and the membranous labyrinth is an oil pipe in the tunnel. The bony labyrinth is full of perilymph; the membranous labyrinth is full of endolymph. The two complex labyrinths are the internal ear (Fig. 3.41).

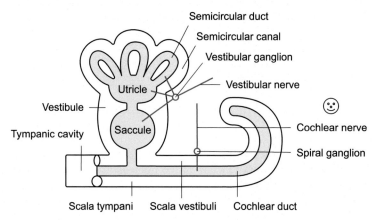

Fig. 3.41   Vestibular nerve from utricle, saccule, semicircular duct; cochlear nerve from cochlear duct.

The bony labyrinth surrounds the membranous labyrinth as follows: the vestibule surrounds the utricle and saccule; the semicircular canal surrounds the semicircular duct (Often, canal is a bony structure and duct is a soft tissue structure.); the scala vestibuli and scala tympani surround the cochlear duct.

Fig. 3.42

During acceleration or deceleration of body shift, endolymph inside the utricle and saccule flows to stimulate a receptor, and the impulse proceeds by way of the vestibular nerve (Fig. 3.41).

Even with one's eyes closed, one knows the inclination of the head. For sensing the inclination in the utricle and saccule, gravity matters rather than inertia.

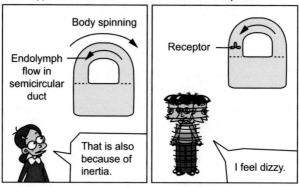

Fig. 3.43

Acceleration or deceleration of body rotation is perceived in the semicircular duct. The impulse is transferred by way of the vestibular nerve too (Fig. 3.41).

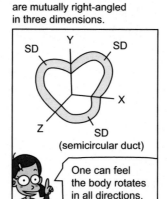

Fig. 3.44

Fig. 3.41 depicts the semicircular ducts in two dimensions, but they actually exist in three dimensions.

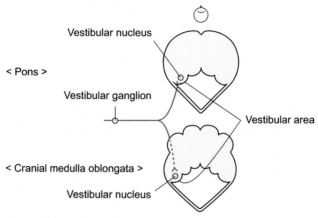

Fig. 3.45 Balance pathway (transverse planes).

The 1st neuron of the "vestibular" nerve arises from the utricle, saccule in "vestibule" and the semicircular duct. The bipolar neuron forms the "vestibular" ganglion (Fig. 3.41) and synapses with the 2nd neuron at the "vestibular" nucleus (Fig. 3.48). The nucleus in the pons and cranial medulla oblongata forms the "vestibular" area (Figs. 1.54, 1.58).

The 2nd neuron from the vestibular nucleus goes to the ventral postero-medial nucleus that is responsible for head sense (Fig. 4.22). The 3rd neuron then proceeds to the vestibular cortex (scattered, not localized) to recognize the various positional changes (Figs. 3.42, 3.43) (Table 2).

Impulse from the vestibular nucleus goes to the vestibulocerebellum as well (Fig. 4.35), to keep balance (Fig. 4.31).

Fig. 3.46

Impulse from the vestibular nucleus also goes to the oculomotor nucleus (Fig. 3.12), trochlear nucleus (Fig. 3.23), and abducens nucleus (Fig. 3.25), to appropriately rotate the eyeballs. This vestibuloocular reflex stabilizes image on the retina (Fig. 3.3) during head movement. This reflex is similar to the spinal cord reflex for knee jerk (Fig. 2.22).

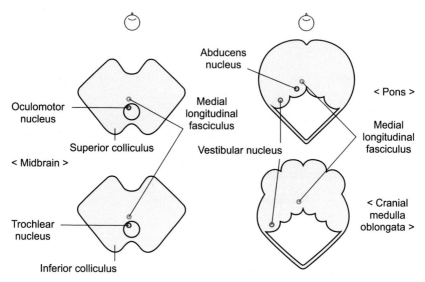

Fig. 3.47 Oculomotor, trochlear, and abducens nuclei, medial longitudinal fasciculus (transverse planes).

Regarding the vestibuloocular reflex, problem is the different levels of the vestibular, oculomotor, trochlear, and abducens nuclei in the brainstem. Solution is the medial longitudinal fasciculus which connects them. As the name implies, the "medial longitudinal fasciculus" resides at "medial" site and runs "longitudinally" and the "fasciculus" is another expression of tract (a bundle of axons). The oculomotor, trochlear, and abducens nuclei are located medially as well (Figs. 3.48, 3.50).

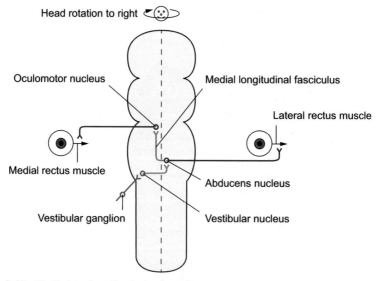

Fig. 3.48 Vestibuloocular reflex by head rotation.

If you rotate your head to the right while watching a fixed object, the neurons transport the impulse as follows. The 1st neuron (involving the vestibular ganglion) from the right semicircular duct (Figs. 3.41, 3.43) synapses with the 2nd neuron at the right vestibular nucleus (Fig. 3.45).

The 2nd neuron decussates and synapses with the 3rd neuron at the left abducens nucleus (Fig. 3.25); the 3rd neuron innervates the left lateral rectus muscle to abduct the left pupil (Fig. 3.13). Some other 3rd neuron decussates and ascends as the medial longitudinal fasciculus (Fig. 3.47). The 3rd neuron synapses with the 4th neuron at the right oculomotor nucleus (Fig. 3.12); the 4th neuron innervates the right medial rectus muscle to adduct the right pupil.

Finally, the bilateral pupils are moved to the left in synchronization, so as to maintain visual aim at the object (Fig. 3.46).

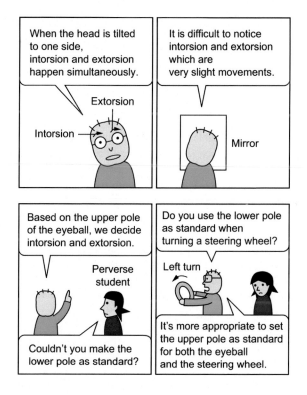

Fig. 3.49

INTorsion and EXTorsion are defined as the INTernal and EXTernal rotations of the eyeball with respect to its anteroposterior axis. This movement results from the vestibuloocular reflex (Fig. 3.46).

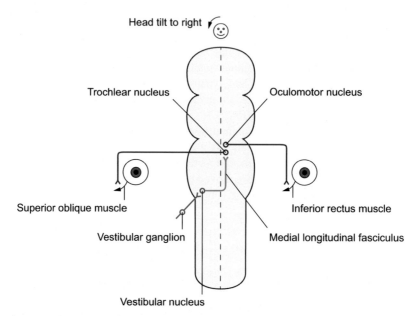

Head tilt to right

Trochlear nucleus

Oculomotor nucleus

Superior oblique muscle

Inferior rectus muscle

Vestibular ganglion

Medial longitudinal fasciculus

Vestibular nucleus

Fig. 3.50   Vestibuloocular reflex by head tilt.

If you tilt head to the right while watching a fixed object, neurons function as follows. The 1st neuron (having the vestibular ganglion) from the right semicircular duct (Figs. 3.41, 3.43) synapses with the 2nd neuron at the right vestibular nuclcus (Fig. 3.45). The 2nd neuron decussates and ascends as the medial longitudinal fasciculus (Fig. 3.47).

The 2nd neuron synapses with the 3rd neuron at the left oculomotor nucleus (Fig. 3.12) that innervates the left inferior rectus muscle to induce extorsion of the left eyeball. The 2nd neuron also synapses with the 3rd neuron at the left trochlear nucleus that decussates and innervates the right superior oblique muscle to induce intorsion of the right eyeball (Fig. 3.13). The lower motor neuron from the trochlear nucleus decussates, which is the exceptional case (Fig. 3.23).

In summary, the medial longitudinal fasciculus in the brainstem is the ascending part of interneuron (green color) that relays impulses from the sensory nerve to the motor nerve (Fig. 3.48). The medial longitudinal fasciculus ascends just after decussation, like the lemniscus of the sensory nerve (Fig. 2.8).

The medial longitudinal fasciculus connecting the oculomotor, trochlear, and abducens nuclei is necessary not only for the unintentional vestibuloocular reflex (Figs. 3.46, 3.49) but also for the intentional ocular movement [e.g., looking left without turning head (Fig. 3.48)].

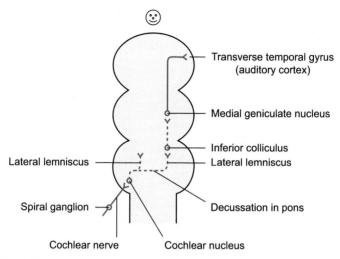

Transverse temporal gyrus (auditory cortex)

Medial geniculate nucleus

Inferior colliculus
Lateral lemniscus

Lateral lemniscus

Decussation in pons

Spiral ganglion

Cochlear nerve   Cochlear nucleus

Fig. 3.51   Auditory pathway.

Regarding the auditory pathway, the 1st neuron of the "cochlear" nerve starts in the "cochlear" duct of internal ear. The 1st neuron is bipolar like that of the vestibular nerve (Fig. 3.41). In such cranial nerves as CN I (Fig. 3.1), CN II (Fig. 3.4), and CN VIII, the 1st neurons are bipolar. Namely, important senses (smell, light, balance sense, and sound) pass the nerve cell body of the 1st neuron (Fig. 2.5).

In the cochlear nerve, the nerve cell bodies of the 1st bipolar neurons are called "spiral" ganglion because they are "spirally" arranged along the cochlear duct, which makes almost three turns unlike Fig. 3.41. [Don't confuse the "spiral" ganglion of the cochlear nerve with the "spinal" ganglion of the spinal nerve (Fig. 3.72).] The 1st neuron ends at the cochlear nucleus.

The 2nd neuron originating from the cochlear nucleus may or may not decussate in the pons; it then ascends as the lateral lemniscus, until it reaches the inferior colliculus of the midbrain (Figs. 1.52, 3.23). As a result, even if the left inferior colliculus is damaged, hearing from the right ear is conveyed to the cerebral cortex. This is similar to the visual pathway, in which the 2nd neuron may or may not decussate (Fig. 3.5). The light and sound are too precious to be missed on either side. Originally, the two eyes and two ears are for the stereoscopic recognition of the light (Fig. 3.6) and sound.

The 3rd neuron extends from the inferior colliculus to the medial geniculate nucleus (Figs. 4.19, 4.48). In succession, the 4th neuron goes to the transverse temporal gyrus (auditory cortex) (Fig. 1.40). It is notable that the auditory pathway leaves the "temporal" bone (Figs. 3.40, 3.41) and arrives in the "temporal" lobe (Fig. 4.11).

If the 2nd and 3rd neurons were united, the auditory pathway would have followed the general rule of afferent nerves having three neurons (Fig. 2.8) (Table 2).

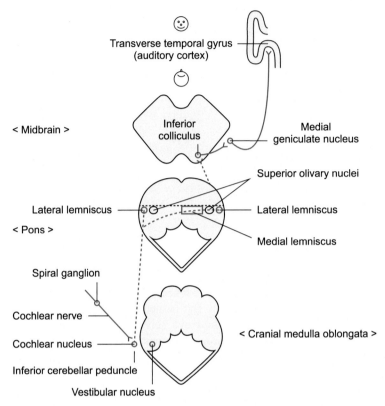

Fig. 3.52 Auditory pathway (transverse planes).

In the transverse planes, the cochlear nucleus resides in the inferior cerebellar peduncle (Fig. 1.54), "lateral" to the vestibular nucleus. (The word "lateral" foreshadows the "lateral" lemniscus.)

The "lateral" lemniscus accompanying the spinal lemniscus (Fig. 2.11) is "lateral" to the medial lemniscus (Fig. 2.12). *(Exactly, before the 2nd neuron becomes the lateral lemniscus, it may form additional synapse at the superior olivary nucleus. Moreover, the 2nd neuron may form additional synapse in the lateral lemniscus.)*

All four lemnisci have been mentioned: spinal lemniscus (Fig. 2.11), medial lemniscus (Fig. 2.12), trigeminal lemniscus (Fig. 3.28), and lateral lemniscus. Commonly, the lemniscus is ascending part of the 2nd neuron of the sensory nerve, after decussation (Tables 1, 2).

The 3rd neuron from the inferior colliculus to the medial geniculate nucleus is identifiable externally (Fig. 4.48). Unlike Fig. 3.51, the 3rd and 4th neurons around the medial "geniculate" nucleus are bent like the neurons around the lateral "geniculate" nucleus (Fig. 3.8).

Cranial nerve IX

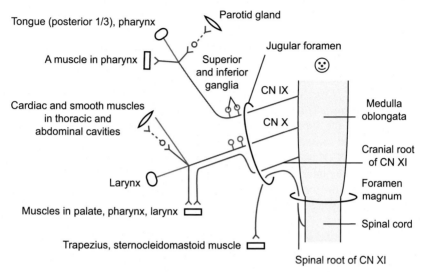

Fig. 3.53   CN IX, X, XI (peripheral nervous system).

CN IX, X, XI are closely related with one another, so the authors call them triple X. The triple X (excluding the spinal root of CN XI) emerges from the retroolivary sulcus of the cranial medulla oblongata (Fig. 1.62). The triple X exits the cranial cavity through the jugular foramen. The internal "jugular" vein is the biggest structure passing through the "jugular" foramen (Fig. 1.21). CN IX, X, XI will be explained one by one with the detailed features of their neurons.

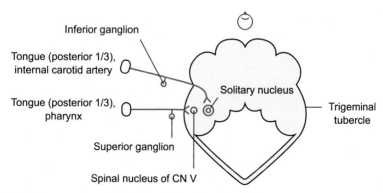

Fig. 3.54   Sensory nerve of CN IX (cranial medulla oblongata).

The spinal nucleus of CN V is shared by CN V (Fig. 3.28), IX, and X (Fig. 3.59). *[Exactly, the spinal nucleus is also shared by CN VII. Accidentally, the spinal nucleus is related to the four cranial nerves in the pharyngeal arches (Fig. 3.36). Nevertheless, the spinal nucleus, related to CN VII, is not introduced in this book because the skin area of CN VII in the auricle is too small.]* Such sharing of nucleus is popular between CN IX and X (Figs. 3.63, 3.68).

Notice that the spinal nucleus of CN V is located inside the trigeminal tubercle (Fig. 3.32). *[Exactly, the trigeminal tubercle is partly covered by the inferior cerebellar peduncle (Figs. 1.51, 3.52).]*

Regarding the superior ganglion of CN IX, the spinal nucleus of CN V receives general sense from "tongue" (posterior 1/3) and "pharynx." Therefore, the name of CN IX is "glossopharyngeal" nerve.

Regarding the inferior ganglion of CN IX, the solitary nucleus receives special sense from the tongue (posterior 1/3). The 2nd and 3rd neurons of the taste pathway are illustrated in Fig. 3.34 (Table 2). *(Exactly, CN X also includes the taste pathway.)*

In addition, the solitary nucleus receives special sense (blood pressure, oxygen concentration) from the internal carotid artery *(exactly, from the carotid sinus and carotid body)*. If blood pressure or oxygen concentration is low in the internal carotid artery, which is in charge of the brain circulation (Fig. 1.3), the solitary nucleus notices it. The solitary nucleus then notifies the cardiovascular center (reticular formation) (Fig. 4.28), which increases heart rate through the sympathetic nerve (Figs. 2.27, 2.31, 3.58).

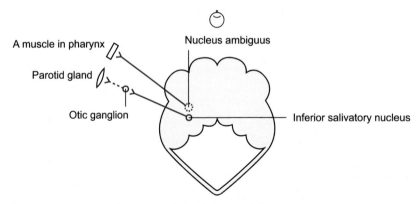

Fig. 3.55   Motor nerve of CN IX (cranial medulla oblongata).

The lower motor neuron from the nucleus ambiguus (meaning unclear nucleus, therefore having unclear boundary) of CN IX terminates in a muscle in pharynx (stylopharyngeus to help swallowing) derived from the 3rd pharyngeal arch (Fig. 3.36).

As a part of the parasympathetic nerve, the inferior salivatory nucleus gives rise to the preganglionic neuron. It goes to the OTIc ganglion for the parOTId gland. (OTI means the ear.) The inferior salivatory nucleus makes pair with the superior salivatory nucleus of CN VII for the two other salivary glands (submandibular and sublingual glands) (Figs. 3.35, 3.37).

Fig. 3.56

The words "salivatory" and "salivary" have the same meaning.

Cranial nerve X

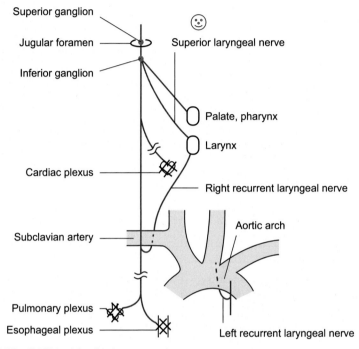

Fig. 3.57 CN X (peripheral nervous system).

This realistic drawing of the complex branches of CN X is explained prior to its neuronal drawings (Figs. 3.59, 3.61). The superior and inferior ganglia of CN X (Fig. 3.53) are part of the sensory nerve.

The 1st branch of CN X innervates the muscles in palate and pharynx (Fig. 3.61). The 2nd branch is the superior laryngeal nerve, which senses the superior part of larynx (Fig. 3.59).

The 3rd branch (parasympathetic nerve) joins the cardiac plexus (Fig. 2.33) to slow down the heart rate. The sympathetic nerve also joins the cardiac plexus (Fig. 2.31) to speed up the heart rate (Fig. 2.27). The cardiac plexus actually resides beside the heart, but is drawn in the neck for figure simplification.

The 4th branch, the recurrent laryngeal nerve, hooks around the subclavian artery (on the right side) or the aortic arch (on the left side). It senses the inferior part of larynx (Fig. 3.59) and controls the muscles in larynx (Fig. 3.61). *(Exactly, a muscle in larynx (cricothyroid muscle) is controlled by the superior laryngeal nerve.)*

The 5th branch joins the pulmonary plexus (Fig. 2.33) to contract the circular smooth muscle in bronchi.

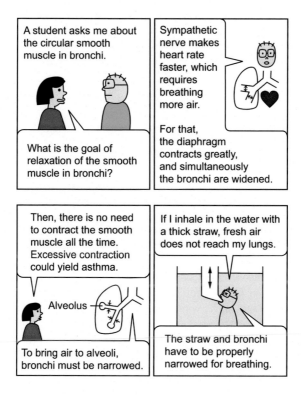

Fig. 3.58

The bronchi are widened by the sympathetic nerve, and narrowed by CN X for fluent respiration.

The last, the 6th branch joins the esophageal plexus and enters the abdominal cavity (Fig. 2.33). In total, CN X innervates a great amount of cardiac and smooth muscles throughout the thoracic and abdominal cavities (Figs. 2.32, 3.57).

Fig. 3.53 depicts the cranial root of CN XI following a rather complicated route. Before passing through the jugular foramen, it is a part of CN XI. However, after passing through the jugular foramen, it is a part of CN X and innervates the muscles in palate, pharynx, and larynx (Fig. 3.61). In neuroanatomy, the cranial root of CN XI is regarded as a part of CN X, taking its function into account.

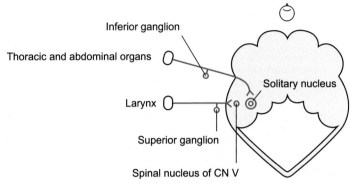

Fig. 3.59 Sensory nerve of CN X (cranial medulla oblongata).

As CN IX contains the superior and inferior ganglia, CN X has its own superior and inferior ganglia (Fig. 3.53). The Superior ganglion is for Somatic sensory nerve, while the Inferior ganglion is for visceral sensory nerve (for Internal organs). For the memorization, imagine that the visceral sensory nerve's huge jurisdiction (thoracic and abdominal organs) pulls the ganglion down, to make it inferior (Fig. 3.57).

The somatic sensory nerve of CN X (from the larynx) (Fig. 3.57) enters the spinal nucleus of CN V. For instance, water mistakenly aspired into the larynx is perceived by CN X, which immediately provokes coughing with the help of other nerves.

The visceral sensory nerve of CN X (from the thoracic and abdominal organs) (Fig. 3.57) enters the solitary nucleus. For instance, the distended stomach is perceived by CN X (Figs. 2.26, 3.60).

Fig. 3.60

In total, the solitary nucleus is for taste (CN VII, IX) (Figs. 3.33, 3.34, 3.54), and for visceral sense of the thoracic and abdominal organs (CN X) (Fig. 3.59).

The solitary nucleus is located in the cranial medulla oblongata (Figs. 3.33, 3.54, 3.59). Its related cranial nerves (CN VII, IX, X) (Fig. 3.68) emerge either from the border between the pons and cranial medulla oblongata or from the cranial medulla oblongata (Fig. 1.62). The internal and external features are associated with each other.

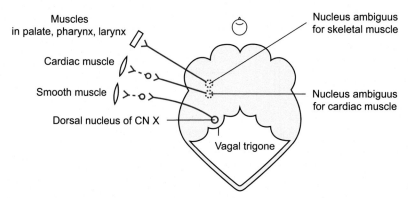

Fig. 3.61   Motor nerve of CN X (cranial medulla oblongata).

The nucleus ambiguus for skeletal muscle (CN X, cranial root of CN XI) is the origin of the somatic motor nerve to the muscles in palate, pharynx, and larynx (Figs. 3.36, 3.53). Recall the nucleus ambiguus of CN IX for a muscle in pharynx (Fig. 3.55).

In addition, the nucleus ambiguus for cardiac muscle (CN X) is the origin of the parasympathetic nerve to the heart (Fig. 3.57), slowing down the heart rate (Fig. 2.27).

Fig. 3.62

The above cartoon may be helpful in memorizing functions of the nucleus ambiguus of CN X.

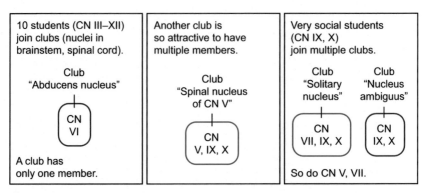

Fig. 3.63

As a rule, each cranial nerve may be related to several nuclei; each nucleus may be related to several cranial nerves. In particular, the spinal nucleus of CN V (Figs. 3.28, 3.54, 3.59), solitary nucleus (Figs. 3.33, 3.54, 3.59), and nucleus ambiguus (Figs. 3.55, 3.61) are related to multiple cranial nerves. The Spinal nucleus of CN V and Solitary nucleus are for Sensory nerve; the nucleus aMbiguus is for Motor nerve (Fig. 3.68).

The dorsal nucleus of CN X sends parasympathetic impulse to the lungs, gastrointestinal tract, and other thoracic and abdominal organs (Figs. 2.32, 3.53, 3.57). Therefore, the dorsal nucleus of "vagus" nerve must be large, to yield the "vagal" trigone in the floor of fourth ventricle (Figs. 1.58, 3.61).

Cranial nerve XI

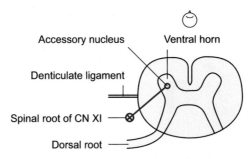

Fig. 3.64 Spinal root of CN XI (spinal cord).

Only the spinal root of CN XI is regarded as the genuine CN XI, excluding the cranial root of CN XI from the nucleus ambiguus (Figs. 3.53, 3.61). The accessory nucleus (official term, nucleus of spinal accessory nerve) is located in the ventral horn of the spinal cord (Fig. 4.45).

The emerging site of the spinal root of CN XI is between the denticulate ligament and the dorsal root of C1−C5 (Fig. 1.68). Namely, the accessory nucleus ends at C5 level, whereas the spinal nucleus of CN V ends at C3 level (Figs. 3.32, 3.68, 5.23).

The spinal root of CN XI enters the cranial cavity through the foramen magnum, and then exits the cranial cavity through the jugular foramen (Fig. 3.53). If the spinal root did not enter the cranial cavity, it would have been considered a spinal nerve.

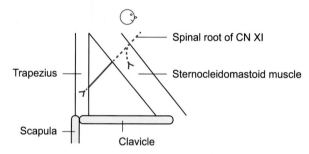

Fig. 3.65 Spinal root of CN XI (peripheral nervous system).

After exiting through the jugular foramen, the spinal root of CN XI inner-vates two muscles (sternocleidomastoid muscle and trapezius) which move the neck (Figs. 4.45, 4.46). *(Exactly, the trapezius extends the neck only when its insertion, scapula is fixed by other muscles.)*

Cranial nerve XII

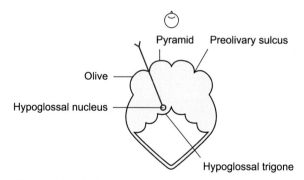

Fig. 3.66 CN XII (cranial medulla oblongata).

In the floor of fourth ventricle, the "hypoglossal" nucleus forms the "hypoglossal" trigone (Fig. 1.58). As the hypoglossal nucleus resides medial to the nuclei of CN IX and X (Fig. 5.22), CN XII emerges [from the preolivary sulcus (ventrolateral sulcus) of the cranial medulla oblongata] medial to CN IX and X (Fig. 1.62).

CN XII controls all the intrinsic and extrinsic muscles of tongue. CN IV (Fig. 3.23), CN VI (Fig. 3.25), spinal root of CN XI (Fig. 3.64), and CN XII contain only somatic motor nerve (Fig. 2.6). These four cranial nerves are even simpler than the spinal nerve (Fig. 3.72).

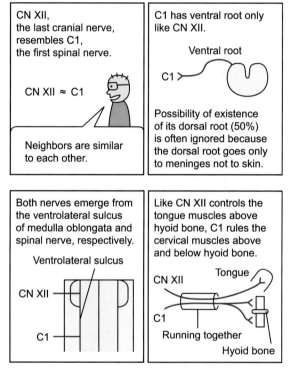

Fig. 3.67

CN XII and C1 are morphologically and functionally similar (Fig. 1.62).

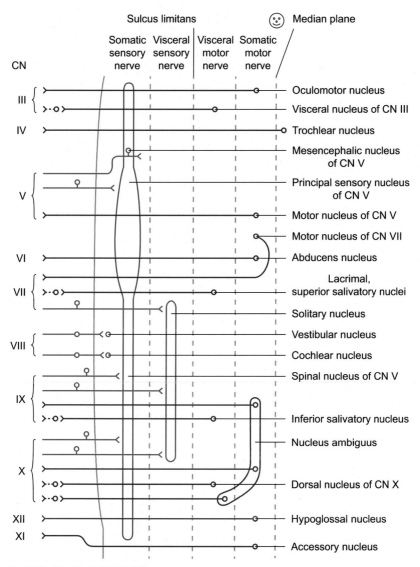

Fig. 3.68   Nuclei of CN III–XII.

In the above figure, the total nuclei of CN III–XII are exhibited according to the level of the midbrain, pons (Its level is represented by the principal sensory nucleus of CN V that is swollen.), medulla oblongata, and spinal cord. The nuclei are also arranged according to the categories of the somatic sensory nerve, visceral sensory nerve, visceral motor nerve, and somatic motor nerve, in terms of the transverse planes of the pons and cranial medulla oblongata (Figs. 5.21, 5.22).

The sensory nuclei of CN V, solitary nucleus, and nucleus ambiguus are long, as if they were continuations of the gray matter of spinal cord (Fig. 1.69). Decussation of the lower motor neuron from the trochlear nucleus is represented (Fig. 3.23). In the peripheral nervous system, the sensory ganglia (Fig. 2.6) and parasympathetic ganglia (Fig. 2.25) are not labeled.

The spinal nerve

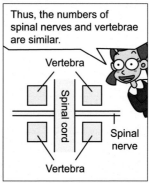

Fig. 3.69

The spinal nerves exit the vertebral canal (Fig. 1.8) by passing through the intervertebral foramina (Fig. 1.66), just like the cranial nerves that exit the cranial cavity (Fig. 1.63).

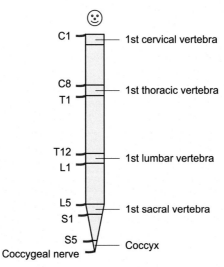

Fig. 3.70 Spinal nerves, vertebrae.

On each side, the spinal nerves are grouped into 8 cervical nerves, 12 thoracic nerves, 5 lumbar nerves, 5 sacral nerves, and 1 coccygeal nerve.

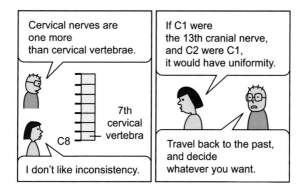

Fig. 3.71

Although C1 seems like the 13th cranial nerve because of its similarity to CN XII (Fig. 3.67), C1 is not regarded as such because of its origin in the spinal cord (Fig. 1.62).

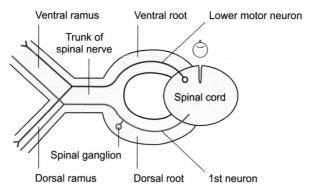

Fig. 3.72 Spinal nerve.

The spinal nerve involves the 1st neuron of the somatic sensory nerve (Fig. 2.8) and the lower motor neuron of the somatic motor nerve (Fig. 2.17). The 1st neuron passes the dorsal root, while the lower motor neuron passes the ventral root (Fig. 2.21). Therefore, the spinal ganglion is found in the dorsal root; its another name is the dorsal root ganglion. The spinal ganglion is located just out of the dura mater (Fig. 1.68) and in the intervertebral foramen (Fig. 3.69).

Regarding the gross anatomy, the dorsal and ventral roots meet to form the trunk of spinal nerve, which immediately divides into the dorsal and ventral rami (Fig. 3.73). Dissimilar to the two roots, the two rami contain both the somatic sensory and somatic motor nerves. The dorsal ramus which is for the deep back muscles is thinner than the ventral ramus which is for most muscles of the trunk and limbs (Figs. 3.73, 3.77, 3.78, 3.81).

Somatic sensory nerve of spinal nerve

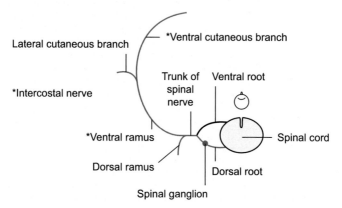

Fig. 3.73 Spinal nerve (T1–T11).

The somatic sensory nerve of the ventral ramus (Fig. 3.72) is discussed, leaving its somatic motor nerve aside. The ventral ramus divides into a lateral cutaneous branch and a ventral cutaneous branch. "Branch" (English) is derived from "ramus" (Latin), as English is derived from Latin.

The ventral rami and ventral cutaneous branches of T1–T11 constitute the intercostal nerves between the adjacent ribs. Their somatic sensory nerve takes charge of cutaneous sense on the thoracic wall. For an instance, the 4th intercostal nerve (T4) receives the sense of nipple on the 4th rib (Fig. 3.74).

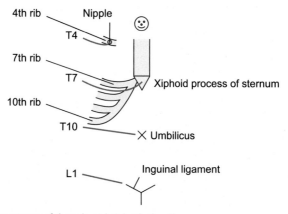

Fig. 3.74 Dermatomes of thoracic and abdominal walls.

Among the intercostal nerves, T7–T11 pass not only the "thoracic" wall, but also the "abdominal" wall. This is the reason T7–T11 are named "thoracoabdominal" nerves. Among them, T10 is distributed to the umbilicus. [Therefore, the umbilicus is represented as X (10 in Roman numerals) in this book]. Adjacent to the subcostal nerve (T12), L1 is distributed to the skin on the inguinal ligament.

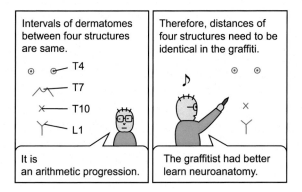

Fig. 3.75

Dermatome is defined as the skin area which is innervated by a spinal nerve. Dermatomes of the nipple (T4), xiphoid process of sternum (T7), umbilicus (T10), and inguinal ligament (L1) have the same intervals between them (Fig. 3.74).

Fig. 3.76

In the fetus (or embryo) posture (Fig. 5.3), dermatomes of upper and lower limbs can be easily recognized and depicted: C5−T1 in the upper limb, L2−S3 in the lower limb (Fig. 1.66).

Somatic motor nerve of spinal nerve

The ventral rami (Fig. 3.72) of C1−C5, those of C5−T1, and those of L2−S3 compose the cervical, brachial, and lumbosacral plexuses, respectively (Fig. 3.82). The brachial and lumbosacral plexuses for abundant limb muscles originate from two enlargements of the spinal cord (Fig. 1.66). What follows is an introduction to the somatic motor nerve of the three plexuses, excluding their somatic sensory nerve (Fig. 3.76).

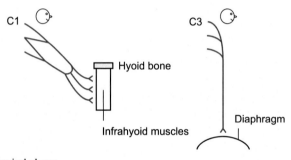

Fig. 3.77   Cervical plexus.

Among the cervical plexus, C1−C3 form a loop, from which branches arise and innervate the infrahyoid muscles. In contrast, the suprahyoid muscles are innervated by CN V (Fig. 3.30) and CN VII (Fig. 3.35). *[Exactly, an infrahyoid muscle (thyrohyoid muscle) and a suprahyoid muscle (geniohyoid muscle) are directly innervated by C1 (Fig. 3.67).]*

C3−C5 form a nerve, which runs downward to reach the diaphragm. In an early developmental stage, the diaphragm *(exactly, a portion of the diaphragm)* has been located cranial to the heart. During head folding (Fig. 5.5), the diaphragm descends to be located caudal to the heart. During this descent, the diaphragm drags along C3−C5 with itself. As a general rule, the nerve follows until the end. The nerves in the pharyngeal arches also exemplify this rule (Fig. 3.36).

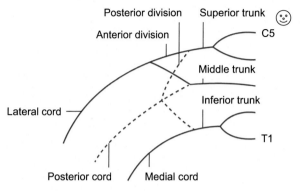

Fig. 3.78 Trunks, divisions, cords of brachial plexus.

In the brachial plexus, the ventral rami of C5—T1 unite and split repeatedly, to form three trunks, six divisions, and three cords. Do you know the old film actor, Robert Taylor? The sentence "Robert Taylor Drinks Coffee" represents the "Rami, Trunks, Divisions, Cords" of brachial plexus. The trunk of brachial plexus differs from the trunk of spinal cord proximal to the ventral ramus (Fig. 3.73).

An important criterion in the brachial plexus is its divisions. Three "anterior" divisions build the lateral and medial cords for the "anterior" muscles, while three "posterior" divisions build the "posterior" cord for the "posterior" muscles.

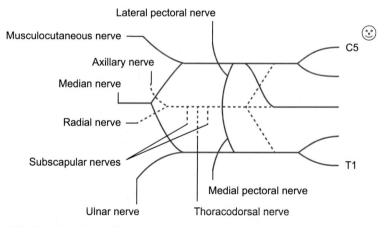

Fig. 3.79 Branches of brachial plexus.

The anterior branches in the above figure are depicted as solid lines. The lateral and medial pectoral nerves go to the pectoral region muscles. The musculocutaneous nerve goes to the anterior arm muscles, while the median and ulnar nerves go to the anterior forearm muscles and palm muscles.

The posterior branches are represented with dotted lines. The subscapular nerves and axillary nerve are for the scapular region muscles; the thoracodorsal nerve is for a superficial back muscle (latissimus dorsi); the radial nerve is for the posterior arm muscles and posterior forearm muscles.

We have trespassed the territory of regional anatomy. Unless the readers belong to the medical field, they do not have to memorize the details of nerves. For these readers, simple understanding of the situation is more than enough.

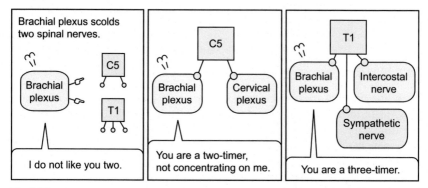

Fig. 3.80

C5 contributes to both the cervical and brachial plexuses (Figs. 3.77, 3.78). Such is common for spinal nerves located on the borderlines (Fig. 3.82).

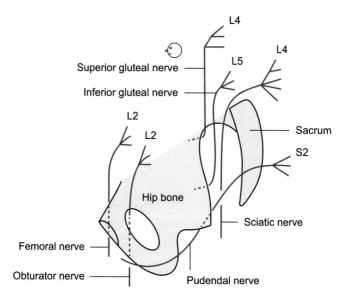

Fig. 3.81   Lumbosacral plexus (medial view).

The brachial plexus innervates muscles in the pectoral region, scapular region, superficial back, and upper limb (Fig. 3.79). Likewise, the lumbosacral plexus innervates muscles in the pelvis, perineum, and lower limb.

In the above figure, branches of the lumbosacral plexus are depicted separately, like branches of the cervical plexus (Fig. 3.77).

The femoral nerve and obturator nerve control the anterior thigh muscles and medial thigh muscles, respectively. Both the superior and inferior gluteal nerves control the gluteal region muscles. These muscles in the thigh and gluteal region are huge in humans walking on two feet (Fig. 2.22).

The sciatic nerve, the thickest nerve in the body, is composed of five spinal nerves (L4−S3); it is noteworthy, considering that the whole brachial plexus is composed of the five spinal nerves (Fig. 3.78). The sciatic nerve is distributed to muscles in vast regions of the lower limb (posterior thigh, whole leg, and whole foot).

The pudendal nerve innervates skeletal muscles in the perineum. Following the Latin word "pudenda" (meaning embarrassing), this nerve is related to defecation, urination, and sexual intercourse. These three activities are performed by the collaboration of skeletal muscle and smooth muscle; the smooth muscle is controlled by the sympathetic and parasympathetic nerves (Fig. 2.36).

Although branches of the lumbosacral plexus are thick for huge muscles in the pelvis, perineum, and lower limb, the branches occupy the small paracentral lobule in somatotopic arrangement (Fig. 4.8).

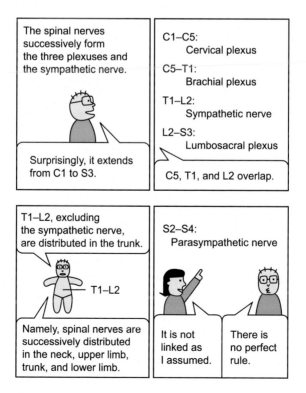

Fig. 3.82

The above cartoon summarizes the whole distribution of the spinal nerves (somatic motor nerve and visceral motor nerve). The overlapping spinal nerves are C5, T1 (Fig. 3.80), and L2. As mentioned, the spinal nerve contains the sympathetic nerve in T1−L2 (Fig. 2.31), and the parasympathetic nerve in S2−S4 (Fig. 2.35).

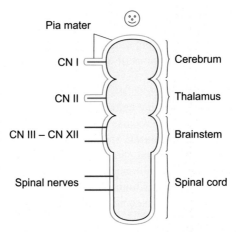

Fig. 3.83 Cranial and spinal nerves from the central nervous system.

In summary of this chapter, CN I from the cerebrum (Fig. 3.1) and CN II from the thalamus (Fig. 3.5) are enclosed by the pia mater (Fig. 1.7). The rest of cranial nerves from the brainstem (exception: the spinal root of CN XI) (Figs. 1.62, 3.68) and the spinal nerves from the spinal cord (Figs. 1.66, 3.72) are not enclosed by the pia mater as the genuine peripheral nervous system (Fig. 1.1.).

Chapter 4

Function of the brain

This chapter explores the comprehensive functions of the brain, which are closely related not only to the morphology of the brain but also to the somatic and autonomic nerves of the cranial and spinal nerves. Details are the functions of the cerebral cortex, limbic system, basal nuclei, diencephalon, cerebellum, and brainstem. Readers can see drawings with consistent style of the brain shape and the neuronal connections. This chapter often refers to the general rule of afferent nerves having three neurons that is developed by the authors. The functions of the brain are explained within the boundary of neuroanatomy; the rest of the functions fall under neurophysiology.

Function of the cerebral cortex

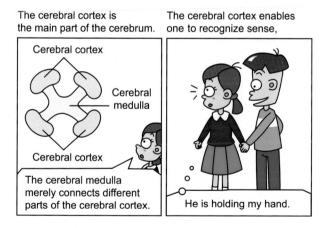

Fig. 4.1

Visually Memorable Neuroanatomy for Beginners. DOI: https://doi.org/10.1016/B978-0-12-819901-5.00004-1

analyze the sense,   and execute motion.

Fig. 4.1   (Continued)

The cerebral cortex plays a key role in all conscious functions of the brain.

Fig. 4.2

Without activity of the cerebral cortex, humans cannot carry out any conscious functions.

Fig. 4.3

Sleeping is a naturally recurring state of the cerebral cortex, characterized by reduced consciousness.

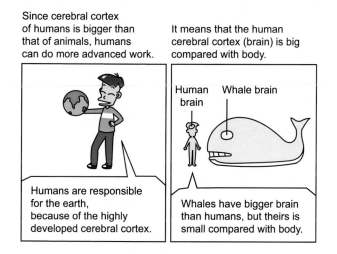

Fig. 4.4

Humans have large brain, mostly due to the size of the cerebral cortex. The human skull has a large cranial cavity (Fig. 1.63). It can be confirmed in a natural history museum that exhibits the skulls of various animals and primitive men.

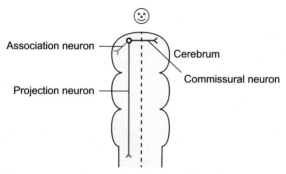

Fig. 4.5 Projection, commissural, and association neurons.

Neurons in the cerebral cortex are organized into three categories. The projection neuron goes to the brainstem or spinal cord. An example is the upper motor neuron of the corticospinal tract before pyramidal decussation (Fig. 2.17). Moreover, the projection neuron involves the sensory pathway (Fig. 2.8), though it is omitted in the above figure.

The commissural neuron goes to the contralateral cerebral hemisphere, through the corpus callosum, anterior and posterior commissures (Fig. 1.44). The association neuron goes to other cerebral cortex in the ipsilateral cerebral hemisphere (Fig. 4.6). In other words, the commissural neuron decussates (Fig. 2.9), while the association neuron does not.

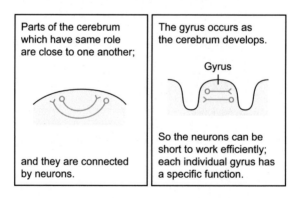

Fig. 4.6

The above cartoon depicts the association neuron in a gyrus. The association neuron also connects the different gyri in a cerebral hemisphere (Fig. 4.5). An example is the neuron running from the speech cortex (inferior frontal gyrus) to the motor cortex (precentral gyrus) (Fig. 4.12).

The above cartoon also tells that sulci and gyri begin to take form on the cerebral hemisphere, which results in the larger cerebral cortex (Fig. 5.9). Simultaneously, each gyrus establishes its specific role.

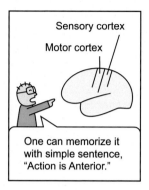

Fig. 4.7

Functionally, the postcentral and precentral gyri are called the sensory cortex (Fig. 2.8) and motor cortex (Fig. 2.17), respectively. In fact, the two cortices include the paracentral lobule on the medial surface of the cerebral hemisphere (Fig. 1.28).

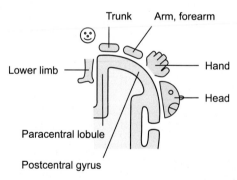

Fig. 4.8 Somatotopic arrangement of sensory cortex.

Somatotopic arrangement is the point-for-point correspondence of the body region to the cerebral cortex through the neuronal pathway.

The best example of the somatotopic arrangement is the medial lemniscus pathway. Fig. 2.14 explains why the upper limb (arm, forearm, hand) and lower limb match the postcentral gyrus and paracentral lobule, respectively. The trunk intervenes between the upper and lower limbs in the cortex. The spinothalamic tract (Fig. 2.11) also follows the same somatotopic arrangement.

128

The somatotopic arrangement also involves the trigeminothalamic tract. Notable fact is that the trigeminothalamic tract passes through the ventral posteromedial nucleus, medial to the ventral posterolateral nucleus (Fig. 4.19). Due to a twist in the corona radiata (Figs. 2.14, 2.16), the head (face, tongue, etc.) correlates with the inferolateral part of postcentral gyrus (Fig. 3.28).

This somatotopic arrangement of the sensory pathways roughly draws a creature on the corresponding gyri. The big-handed and big-headed creature is called sensory homunculus.

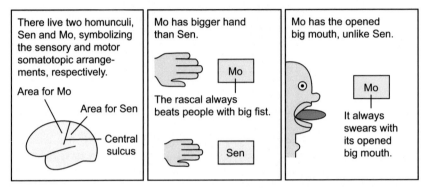

Fig. 4.9

The motor homunculus resides in the precentral gyrus and paracentral lobule, and differs slightly from the sensory homunculus.

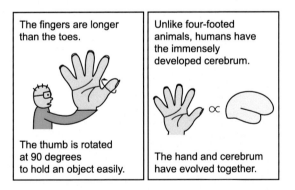

Fig. 4.10

Both the sensory and motor homunculi are depicted with a big hand (Figs. 4.8, 4.9). It implies that during evolution, the human cerebrum has enlarged (Fig. 4.4) with free hand movement. Highly developed cerebrum and hand movement are closely related. For example, to memorize information, a student draws and writes on paper with hand (Fig. 4.33), activating large area of the cerebral cortex.

In contrast, both the sensory and motor homunculi possess the very small lower limb (Fig. 4.8) in spite of the large muscles and thick nerves (Fig. 3.81) in it. So it is hard to use tools that require fine movement (such as smartphone) with the lower limb.

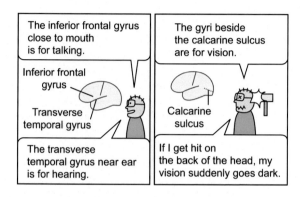

Fig. 4.11

The frontal lobe (left cerebral hemisphere), temporal lobe, and occipital lobe have gyri for talking (speech cortex) (Fig. 4.12), hearing (auditory cortex) (Fig. 3.52), and seeing (visual cortex) (Fig. 3.10), respectively. The above cartoon explains their locations ridiculously. Basically, the frontal lobe deals with the motor nerve, while the rest of the lobes deal with the sensory nerve (Figs. 4.25, 5.15). An exception is the insula (Fig. 1.27) and the opercular part of the frontal lobe (Fig. 1.26) for the taste sense (Fig. 3.34).

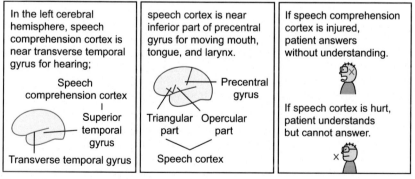

Fig. 4.12

130

The speech comprehension cortex (Wernicke area) converts speech to words; the speech cortex (Broca area) converts words to speech. The above cartoon explains the locations (Fig. 1.26) and impairments of these cortices. The Left cerebral hemisphere is in charge of Language.

Function of the limbic system

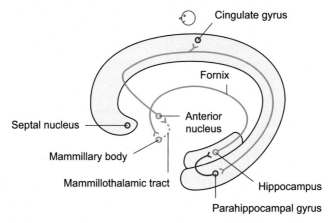

Fig. 4.13   Medial limbic circuit (medial view).

The limbic system is close to the "medial" surface of the cerebral hemisphere (Figs. 1.28, 1.30, 1.44). That is why neuronal circuit in the limbic system is named the "medial" limbic circuit (Papez circuit).

The limbic system roughly follows the general rule of afferent nerves (Table 3). The 1st neuron starts at the hippocampus (center of limbic system), curves around as the fornix (Fig. 1.35), and synapses with the 2nd neuron at the mammillary body (Figs. 1.44, 1.62). The 2nd neuron drawn as dotted line does not decussate (not following the rule), and synapses with the 3rd neuron at the anterior nucleus of thalamus (following the rule) (Fig. 4.19); the 2nd neuron is called the mammillothalamic tract (official term, mammillothalamic fasciculus). The 3rd neuron goes to the cerebral cortex of the cingulate gyrus and parahippocampal gyrus (following the rule) (Fig. 1.28).

Next, the impulse in the parahippocampal gyrus goes back to the hippocampus (Fig. 1.30). This neuron may be regarded as the preliminary neuron before the 1st neuron. Together, the neurons configurate the medial limbic circuit.

The limbic system receives sensory impulses indirectly, except the olfactory impulse which is received directly (through the amygdaloid nucleus, a part of limbic system) (Figs. 1.39, 3.1). The limbic system processes the sensory impulses to carry out two functions: memory and emotion.

Memory is stored in the hippocampus. For instance, an animal's memory of dangerous predator is stored in the hippocampus for self-protection. The smell of predator sensed by the CN I (Fig. 3.1) directly stimulates the limbic system to provoke emotion of fear.

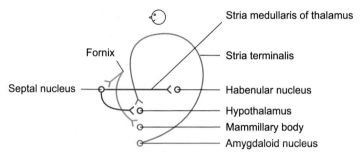

Fig. 4.14   Limbic system influencing epithalamus, hypothalamus.

Regarding the limbic system's influence, a part of the fornix arrives at the septal nucleus (Figs. 1.28, 4.13). The impulse in the septal nucleus reaches the epithalamus, influencing it (stria medullaris of thalamus, habenular nucleus) (Figs. 1.44, 1.45).

The limbic system influences the hypothalamus as follows. According to the medial limbic circuit, the fornix sends an impulse to the mammillary body, a part of hypothalamus (Fig. 4.13). Another impulse from the septal nucleus goes to the hypothalamus.

The other impulse from the amygdaloid nucleus (Fig. 1.39) to the hypothalamus is conveyed along the stria terminalis, which accompanies the caudate nucleus (Fig. 1.40). (If the amygdaloid nucleus is removed, the animal does not show emotion of fear.) The "stria" medullaris of thalamus and the "stria" terminalis belong to the epithalamus and the limbic system, respectively.

All impulses from the limbic system affect the hypothalamus' hormone secretion and autonomic nerve regulation (Figs. 4.27, 4.28, 4.29). The limbic system and hypothalamus work for the past experience (memory and emotion) and new adaptation, in sequence.

Fig. 4.15

While the limbic system keeps memory of the past, the brainstem, influenced by the hypothalamus (Fig. 4.28), deals with the present (Fig. 4.43), and the whole cerebrum predicts the future. The evolution of the brainstem was completed firstly, the limbic system secondly, and the cerebrum thirdly. The more recent a structure's finalization of evolution, the more complex its function and morphology.

The nerve that has primitive function has primitive histological structure. That is why the hippocampus (CA1, CA2, CA3, CA4) (Fig. 1.33) is structured more simply (three or four layers) than the typical cerebral cortex (six layers).

Function of the basal nuclei

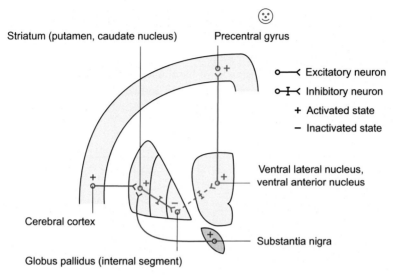

Fig. 4.16   Direct pathway of basal nuclei.

Again, keep the general rule of afferent nerves in mind (Table 3). Just as the hippocampus is the center of limbic system (Fig. 4.13), the striatum (made up of the putamen and caudate nucleus) is the center of basal nuclei (Fig. 1.37). Just as the 1st neuron begins at the hippocampus, the 1st neuron begins at the striatum.

The striatum receives the preliminary neuron from the cerebral cortex, which intends a motion. In order for the striatum to communicate with the large area of cerebral cortex, the striatum is located at the basal area of the cerebrum (Fig. 1.38), and its caudate nucleus has been elongated (Fig. 1.39). The striatum also receives the preliminary neuron from the substantia nigra *(exactly, compact part of substantia nigra)*, which is a gray matter in the midbrain (Fig. 1.52).

The 1st neuron from the striatum arrives at the globus pallidus (internal segment). *(Exactly, the internal segment shares its role with the reticular part of substantia nigra.)*

Plenty of neurons go to and come from the striatum, the center of basal nuclei. The neurons form "striations" (stripes) in the "striatum" (putamen, caudate nucleus) which are roughly visible in the brain slices and brain MRIs (Fig. 1.40).

Then the 2nd neuron from the globus pallidus (internal segment) does not decussate, but goes to the ventral lateral and ventral anterior nuclei of thalamus (Fig. 4.19). The 3rd neuron reaches the precentral gyrus, the motor cortex (Fig. 2.17). The neurons tend to follow the general rule of afferent nerves (Table 3). *(Exactly, the 3rd neuron goes to the frontal lobe and eventually influences the precentral gyrus.)*

This is called the direct pathway of basal nuclei. The direct pathway activates the precentral gyrus because the 1st and 2nd neurons are inhibitory. The negative of the negative equals the positive. $(-1) \times (-1) = (+1)$. Be aware of the fact that ordinary neurons are excitatory.

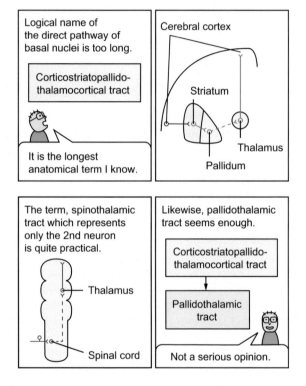

Fig. 4.17

The direct pathway is termed the corticostriatopallidothalamocortical tract.

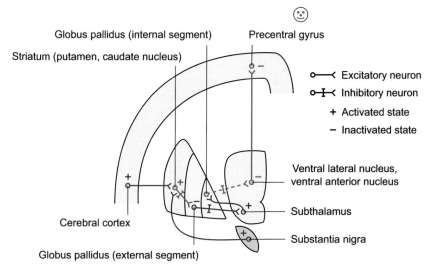

Fig. 4.18   Indirect pathway of basal nuclei.

The other is the indirect pathway, which additionally includes neurons between the globus pallidus and subthalamus (Fig. 4.26). Topographically, the SUBthalamus occupies the inferoposterior area of the diencephalon to be in close contact with the SUBstantia nigra of the midbrain (Fig. 1.52).

A simple mnemonic: In both direct and indirect pathways, the "internal" segment sends impulse to the thalamus, the most "internal" structure of the pathways. The neurons connecting the two structures (pallidothalamic tract) are drawn as dotted lines in Figs. 4.16, 4.17, 4.18.

Along the indirect pathway, the 1st neuron to the globus pallidus (external segment) is inhibitory. The 2nd neuron to the subthalamus is inhibitory, while the reversing 3rd neuron to the globus pallidus (internal segment) is excitatory. The 4th neuron to the thalamus is inhibitory, like in the direct pathway (Fig. 4.16). As a result, the INdirect pathway INactivates the precentral gyrus. $(-1) \times (-1) \times (-1) = (-1)$. Compared with the direct pathway, the SUBThalamus adds SUBTraction (minus sign).

Let us summarize two pathways. In the direct pathway, the 1st and 2nd neurons are inhibitory (Fig. 4.16). In the indirect pathway, the 1st and 2nd neurons are also inhibitory; so is the 4th neuron. Only for memorization of the ordinal numbers, following advanced equations are suggested: $(-1) \times (-2) = (+2)$. $(-1) \times (-2) \times (-4) = (-8)$.

Consequently, the direct and indirect pathways of the basal nuclei alternatively activate and inactivate the precentral gyrus to yield appropriate movement (not too little, not too much) (Fig. 4.23).

Function of the diencephalon

It is suggested to read this subchapter after reading the rest of this chapter "Function of the brain" because the diencephalon works with all other parts of the brain. Main of the diencephalon is the thalamus, where the 2nd neurons of afferent nerves arrive, and the 3rd neurons launch (Fig. 2.8).

The thalamus (Fig. 1.43) is like the CEO (cerebrum)'s secretary (Figs. 2.18, 4.25). The olfactory pathway is like the CEO's family who does not have to go through the secretary (Fig. 3.1).

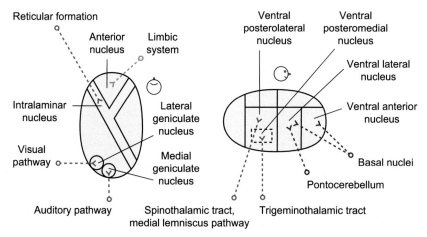

Fig. 4.19 Afferent nerves to thalamic nuclei.

The thalamic nuclei, according to the afferent nerves, are summarized in the above figure and Tables 1, 2, 3. Detailed explanations are provided below.

Fig. 4.20

The intralaminar nucleus is the station where the ascending reticular activating system passes (Fig. 4.42).

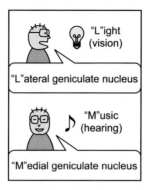

Fig. 4.21

The lateral and medial geniculate nuclei receive the visual pathway (Fig. 3.5) and auditory pathway (Fig. 3.51), respectively.

Fig. 4.22

The ventral posterolateral nucleus is assigned to the spinal nerve (spinothalamic tract, medial lemniscus pathway) (Fig. 2.8); the ventral posteromedial nucleus is assigned to CN V (trigeminothalamic tract) (Fig. 3.28), CN VII, IX (taste pathway) (Fig. 3.34), and CN VIII (balance pathway) (Fig. 3.45).

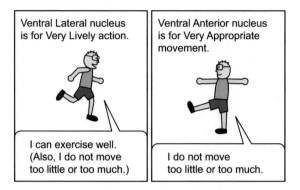

Fig. 4.23

The ventral lateral nucleus is assigned to the pontocerebellum (Fig. 4.37); the ventral lateral nucleus and ventral anterior nucleus are assigned to the basal nuclei (direct and indirect pathways) (Figs. 4.16, 4.18).

Fig. 4.24

The anterior nucleus is assigned to the limbic system (mammillothalamic tract) (Fig. 4.13).

Excluding the intralaminar and geniculate nuclei, the more "cranial" (anterior) a thalamic nucleus is, the more "cranial" the related structure (or pathway) is (Fig. 4.19) (Tables 1, 2, 3).

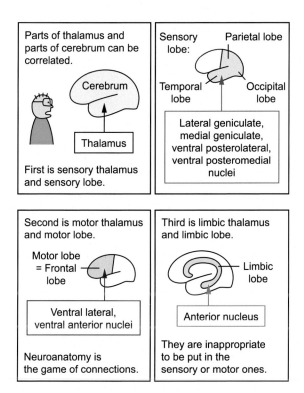

Fig. 4.25

The 3rd neurons from the thalamus go to the allocated parts of the cerebral cortex: Neurons from most nuclei (sensory pathways) to the parietal, occipital, and temporal lobes; neurons from the ventral lateral and ventral anterior nuclei (pontocerebellum, basal nuclei) to the frontal lobe (precentral gyrus) (Figs. 4.16, 4.18, 4.37); neurons from the anterior nucleus (limbic system) to the cingulate and parahippocampal gyri (Figs. 4.13, 4.19) (Tables 1, 2, 3).

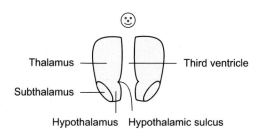

Fig. 4.26   Hypothalamus, adjacent structures.

Beneath the thalamus, is the hypothalamus, which is identifiable in the third ventricle (Fig. 1.44). Like the thalamus (Fig. 1.43), the hypothalamus is constituted by many nuclei, which are not detailed in this book.

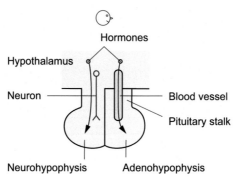

Fig. 4.27   Hypothalamus, pituitary gland.

The first of the hypothalamus' duties is endocrine function. Through the neuron and blood vessel, hormones are conveyed from the hypothalamus to the neurohypophysis on posterior side and adenohypophysis on anterior side, respectively (Fig. 5.13).

Suppose the endocrine system is an athletic team. The pituitary gland (captain of athletes) is influenced by the hypothalamus (coach of team), and consequently influences other glands (other athletes). The team play of the endocrine system is essential for adaptation to the changing environment.

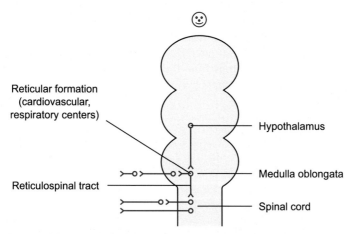

Fig. 4.28   Hypothalamus, reticular formation influencing motor nerves.

The second of the hypothalamus' duties is autonomic nerve function. The hypothalamus controls the visceral motor nerve (Fig. 2.25) and some somatic motor nerve (Fig. 2.6) (in the cranial and spinal nerves) by way of the reticular formation (Fig. 4.41).

Examples of such reticular formation are the cardiovascular center and respiratory center (Fig. 4.43) which are found in the medulla oblongata. The two centers are in relation to the cardiac muscle (Figs. 2.31, 2.33) and the diaphragm. The diaphragm (Fig. 3.77) is controlled both by the corticospinal tract (Fig. 2.17) and the reticulospinal tract. So, you can breathe voluntarily when think about it and involuntarily when not think about it.

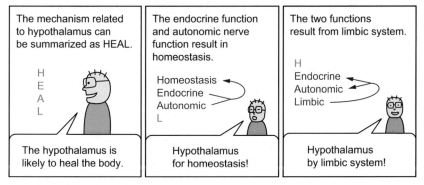

Fig. 4.29

Finally, the Hypothalamus keeps Homeostasis via the endocrine system and autonomic nerve. A difference is that the endocrine system generates slower effects than the autonomic nerve. The hypothalamus is affected by the limbic system (Fig. 4.14).

Function of the cerebellum

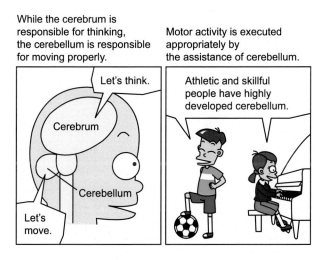

Fig. 4.30

The cerebellum ensures the synchronized contractions of different muscles during movement, by accumulating and releasing the movement-related information through neuronal connection. This advisory function is undertaken by three parts of the cerebellum.

The vestibulocerebellum enables one to keep balance.

Fig. 4.31

The vestibulocerebellum (flocculonodular lobe) (Fig. 1.50) contributes to balance. The VESTIBULocerebellum is related with the VESTIBULe of the internal ear (Figs. 3.41, 4.35), perceiving body movement and posture.

The spinocerebellum enables one to contract muscle with appropriate force.

Fig. 4.32

The spinocerebellum (Fig. 1.50) contributes to suitable force. The SPINocerebellum receives impulse from the SPINal nerve (Fig. 4.36).

Fig. 4.33

The pontocerebellum (Fig. 1.50) contributes to intentional skilled movement such as writing. The PONtocerebellum is connected to the PONs (Fig. 4.37).

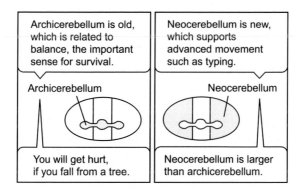

Fig. 4.34

Regarding evolution, the vestibulocerebellum (archicerebellum) is very old; the spinocerebellum ($\fallingdotseq$ paleocerebellum) is old; the pontocerebellum ($\fallingdotseq$ neocerebellum) is new (Fig. 1.50). In the nervous system, the evolutionally new structures are big enough to accommodate complicated functions.

144

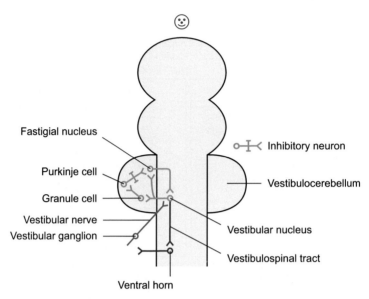

Fastigial nucleus

Inhibitory neuron

Purkinje cell

Vestibulocerebellum

Granule cell

Vestibular nerve

Vestibular ganglion

Vestibular nucleus

Vestibulospinal tract

Ventral horn

Fig. 4.35   Pathway of vestibulocerebellum.

The cerebellar hemisphere is depicted as a semicircle beside the brain-stem (Figs. 2.8, 4.40). All the vestibulocerebellum, spinocerebellum, and pontocerebellum (Fig. 1.50) contain a loop of neurons that coordinate muscle contraction. Each loop is depicted in green (Figs. 4.36, 4.37).

Concerning the vestibulocerebellum, the vestibular nerve from the vestibular ganglion (Fig. 3.41) synapses at the vestibular nucleus (Fig. 3.45). The loop starts here, passing the granule cell, Purkinje cell, and fastigial nucleus consecutively (Fig. 1.50).

The granule cell and Purkinje cell are localized in the cerebellar cortex (Fig. 1.47). However, the loop does not generate conscious recognition which is possible only in the cerebral cortex (Fig. 4.1).

The fastigial nucleus (a kind of cerebellar nucleus) is the leader of the vestibulocerebellum (Fig. 1.50). The Purkinje cell is an inhibitory neuron that influences the fastigial nucleus negatively. For equilibrium, the fastigial nucleus requires an excitatory neuron from the vestibular nucleus. This excitatory shortcut also appears in the spinocerebellum and pontocerebellum, where the emboliform, globose nuclei and dentate nucleus (cerebellar nuclei) act as their leaders, respectively (Figs. 4.36, 4.37).

After the loop, motor neuron from the vestibular nucleus descends to the ventral horn; this neuron is the vestibulospinal tract. During a gymnast's performance on a balance beam, the vestibular nerve sends the balance sense to the vestibulocerebellum, which advises the body not to fall off (Fig. 4.31).

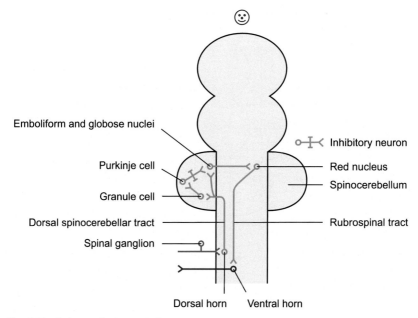

Fig. 4.36 Pathway of spinocerebellum.

With regard to the spinocerebellum, the 1st neuron synapses at the dorsal horn; the 2nd neuron named Dorsal spinocerebellar tract Directly arrives at the ipsilateral spinocerebellum (Fig. 1.50). Meanwhile, the Ventral spinocerebellar tract decussates twice (Via the midbrain) to reach the ipsilateral spinocerebellum (Fig. 4.38).

The open loop of the spinocerebellum passes the granule cell, Purkinje cell, emboliform and globose nuclei (Fig. 1.50), and contralateral red nucleus (Fig. 1.52).

Motor neuron from the red nucleus decussates and descends to the ventral horn; this neuron is the rubrospinal tract. *[Exactly, not only the red nucleus and rubrospinal tract, but also the (ipsilateral) reticular formation (Fig. 4.41) and reticulospinal tract (Fig. 4.28) are involved.]* When gripping an object, the spinal nerve sends the proprioception (object's firmness, etc.) to the spinocerebellum, which in return advises the hand and forearm to grip the object with appropriate force (Fig. 4.32).

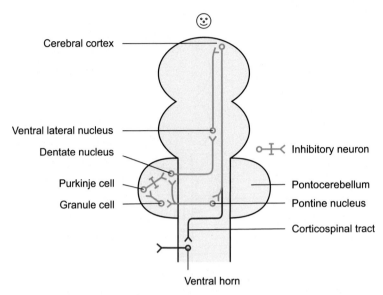

Fig. 4.37 Pathway of pontocerebellum.

The loop of the pontocerebellum starts from the cerebral cortex and passes the pontine nucleus (Fig. 1.54), granule cell, Purkinje cell, dentate nucleus (Fig. 1.50), and ventral lateral nucleus (Fig. 4.19).

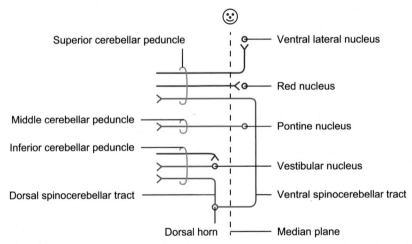

Fig. 4.38 Afferent and efferent nerves passing through superior, middle, and inferior cerebellar peduncles.

The above figure demonstrates the afferent nerve (blue) and efferent nerve (red) of the cerebellum passing through one of the superior, middle, and inferior cerebellar peduncles (Fig. 1.48). The passage is congruent with the afferent and efferent nerves' affiliations (midbrain, pons, medulla oblongata, etc.) (Fig. 1.51).

In the vestibulocerebellum, the afferent and efferent nerves are connected to the vestibular nucleus (pons and medulla oblongata) (Fig. 3.45). It is the inferior cerebellar peduncle where both the afferent and efferent nerves pass, because the vestibular nucleus is mainly located at the medulla oblongata (Fig. 3.52).

In the spinocerebellum, an afferent nerve (dorsal spinocerebellar tract) from the dorsal horn (spinal cord) passes through the inferior cerebellar peduncle, while the other afferent nerve (ventral spinocerebellar tract) passes through the superior cerebellar peduncle. The efferent nerve going to the red nucleus (midbrain) (Fig. 1.52) passes through the superior cerebellar peduncle.

To elaborate on the ventral spinocerebellar tract, the 1st neuron synapses at the dorsal horn and the 2nd neuron immediately decussates, as in the spinothalamic tract (Fig. 2.11). [This is unexpected anatomy because the spinocerebellar tract mainly conveys proprioception like the medial lemniscus pathway (Fig. 2.12).]

In the pontocerebellum, the afferent nerve coming from the pontine nucleus (Fig. 1.54) passes through the middle cerebellar peduncle. The efferent nerve going to the ventral lateral nucleus (thalamus) (Fig. 4.19) passes through the superior cerebellar peduncle.

Returning to Fig. 4.37, impulse from the cerebral cortex to the pontocerebellum is huge, resulting in the huge size of the pontine nucleus (Fig. 1.54), middle cerebellar peduncle (Fig. 1.48), pontocerebellum, and dentate nucleus (Fig. 1.50).

After the loop of pontocerebellum, neuron from the cerebral cortex (in detail, precentral gyrus) descends to the ventral horn, which is the upper motor neuron of the corticospinal tract (Figs. 2.17, 4.37). When we intend to write, neuron from the cerebral cortex sends the intention to the pontocerebellum, which instructs harmonious contraction of the involved muscles (Fig. 4.33).

The pontocerebellum includes two decussations in the loop, so the left cerebral cortex is linked with the right pontocerebellum. The corticospinal tract's upper motor neuron decussates, so the left cerebral cortex is linked with the right ventral horn. Consequently, if the right pontocerebellum is damaged, the patient is unable to write well with the right hand (Fig. 4.37). Likewise, the right vestibulocerebellum and right spinocerebellum corresponds with the right side of the body (Figs. 4.35, 4.36).

In the pontocerebellum (Fig. 4.37), let's focus on the afferent nerve that runs (from the cerebellar cortex) to the cerebral cortex. The 1st neuron (Purkinje cell) synapses with the 2nd neuron at the dentate nucleus (Fig. 1.50). The 2nd neuron decussates and synapses with the 3rd neuron at the ventral lateral nucleus of thalamus (Fig. 4.19). The 3rd neuron arrives at the cerebral cortex. The three neurons follow the general rule of afferent nerves (Table 3). *(Exactly, the three neurons to the cerebral cortex exist in the vestibulocerebellum and spinocerebellum as well.)*

Let's discuss about the afferent nerve to the cerebellar cortex. In all the vestibulocerebellum, spinocerebellum, and pontocerebellum, three successive neurons are required to approach the Purkinje cell. Unexpectedly, the 3rd neuron (granule cell) does not originate from the thalamus (Figs. 4.35, 4.36, 4.37). Therefore, the three neurons are not worth being listed in Table 3.

148

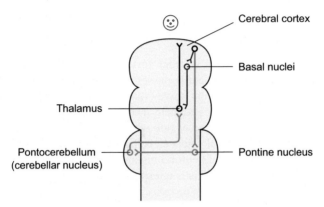

Fig. 4.39 Pathways of basal nuclei (purple), pontocerebellum (green).

The basal nuclei and cerebellar nuclei are morphologically equivalent because they are deep in the cerebral medulla (Figs. 1.40, 5.11) and cerebellar medulla (Fig. 1.47).

Both the basal nuclei and pontocerebellum (containing cerebellar nucleus) receive impulse from the cerebral cortex, and send impulse back to the cerebral cortex by way of the thalamus. The thalamus for basal nuclei is the ventral lateral and ventral anterior nuclei (Figs. 4.16, 4.18), whereas that for pontocerebeLLum is the ventraL Lateral nucleus (Figs. 4.19, 4.37) (Table 3).

All the basal nuclei (Figs. 4.16, 4.18), vestibulocerebellum (Fig. 4.35), spinocerebellum (Fig. 4.36), and pontocerebellum (Fig. 4.37) influence the lower motor neuron not directly, but via the cerebral cortex or brainstem. Contrastingly, the basal nuclei minimize unintentional movement, while the cerebellum enhances intentional movement.

Function of the brainstem

Since the brainstem connects the cerebrum, cerebellum, and spinal cord, plenty of motor and sensory nerves pass through it.

Therefore, if the brainstem is damaged, motor and sensory activities become severely restricted.

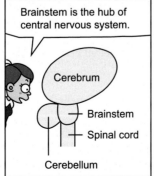

Fig. 4.40

The brainstem works as a relay of the somatic sensory nerve (Fig. 2.8) and somatic motor nerve (Fig. 2.17). The brainstem contains the nuclei and tracts of CN III−XII to support their various activities (Figs. 1.62, 3.68). In this subchapter, the rest of the brainstem functions are discussed.

Reticular formation

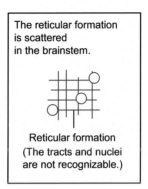

Fig. 4.41

The "reticular" formation is a "network" composed of vague tracts and nuclei (Fig. 2.4) in the brainstem. This primitive structure is evolutionally old and also present in animals, so it is assumed that the reticular formation works for lower level activity, namely for survival.

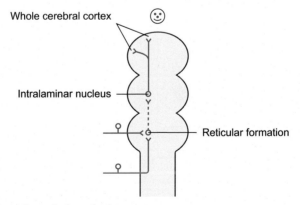

Fig. 4.42 Ascending reticular activating system.

Including the reticular formation, the ascending reticular activating system is responsible for enhancing consciousness (Fig. 4.3). This system responds to all kinds of external stimuli. An example is waking up by the sound of alarm clock (Fig. 4.20).

In this ascending reticular activating system, the 1st neuron in the cranial and spinal nerves synapses with the 2nd neuron in the reticular formation. The 2nd neuron ascends and synapses with the 3rd neuron at the intralaminar nucleus of thalamus (Fig. 4.19). The 3rd neuron diffusely ends at the whole cerebral cortex (Table 3).

Since the brainstem is responsible for circulation and respiration, one cannot survive if it is damaged.

Simply put, removal of even a very small amount of brainstem can cause death.

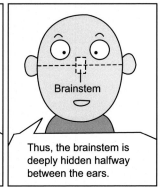

No one can survive if her/his heart stops beating and she/he doesn't breathe.

Thus, the brainstem is deeply hidden halfway between the ears.

On the other hand, one can still survive if a good portion of the cerebrum is removed.

But one will become mentally retarded.

What is this figure?

Fig. 4.43

Parts of the reticular formation (cardiovascular and respiratory centers) in the medulla oblongata control heartbeat and breathing (Fig. 4.28).

In case of brainstem death, the patient shows no spontaneous breathing.

In this case, the organs can be transplanted to other patients.

Respirator

Alive organ

Brainstem death

In many cases, other organs are still alive.

It is the noblest way to use the patient's organs.

Fig. 4.44

In case of cerebrum death (vegetative state), the patient has no consciousness (Fig. 4.2). In case of brainstem death (brain death), the patient has neither consciousness nor self-breathing (Fig. 4.43).

Superior colliculus

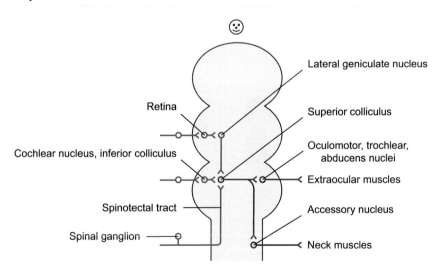

Lateral geniculate nucleus

Retina

Superior colliculus

Cochlear nucleus, inferior colliculus

Oculomotor, trochlear, abducens nuclei

Extraocular muscles

Spinotectal tract

Accessory nucleus

Spinal ganglion

Neck muscles

Fig. 4.45 Pathway of superior colliculus.

The superior colliculus in the midbrain (Fig. 1.52) is the reflex center of eyeballs and neck, regarding external stimuli. In the pathway of reticular formation (Figs. 4.28, 4.42) and the pathway of superior colliculus, do not mind the inconsistent decussation. They are primitive and not well-organized.

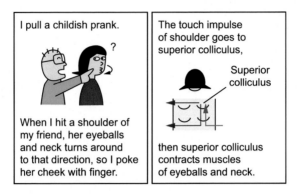

Fig. 4.46

Touch is an impulse that ascends from the spinal nerve to the superior colliculus through the spinotectal tract. [The superior colliculus belongs to the tectum (Fig. 1.52).] Then, in the superior colliculus, the upper motor neuron is initiated. It influences the oculomotor nucleus (Fig. 3.12), trochlear nucleus (Fig. 3.23), abducens nucleus (Fig. 3.25) in the brainstem, and the accessory nucleus (Fig. 3.64) in the spinal cord, causing rotation of the eyeballs (Fig. 3.13) and neck (Figs. 3.65, 4.45).

Fig. 4.47

Sound is another impulse that ascends from the inferior colliculus (Fig. 3.52) to the nearby superior colliculus. Light is the other impulse that proceeds from the lateral geniculate nucleus (Fig. 3.5) to the superior colliculus (Fig. 4.48). The sound and light also result in the rotation of the eyeballs and neck (Fig. 4.45).

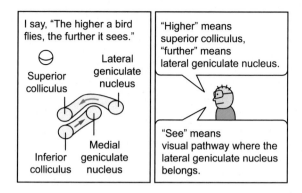

Fig. 4.48

The superior colliculus is connected with the lateral geniculate nucleus for reflex (Fig. 4.45), while the inferior colliculus is connected with the medial geniculate nucleus for auditory pathway (Fig. 3.52). The connection between the lateral geniculate nucleus and superior colliculus is also for light reflex, because the connection includes the 2nd neuron of visual pathway (Fig. 3.5) to the pretectal nucleus (Fig. 3.18).

With a brain specimen, the two connections (official terms, brachia of superior and inferior colliculi) are to be recognized in dorsal view of the midbrain (Fig. 1.52), under the pulvinar (Figs. 1.43, 1.45).

While the reticular formation enhances consciousness by the external stimuli (Fig. 4.42), the superior colliculus enhances attention by the external stimuli (Fig. 4.45).

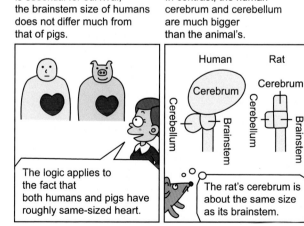

Fig. 4.49

Unlike the cerebrum (Fig. 4.4) and cerebellum (Fig. 4.33), the brainstem of humans is not large compared with that of animals. In other words, the brainstem function of humans and that of animals do not differ that much.

Chapter 5

Development of the central nervous system

Neuroanatomy and development of the central nervous system are like the result and its cause, respectively. Therefore, neuroanatomy can be well understood by knowing development. Reversely, development can be understood after knowing neuroanatomy, so this chapter is the book's last part. The first form of the brain and spinal cord is the neural tube, which originates from the ectoderm. The five brain vesicles of the neural tube then develop and flex to become the cerebrum, diencephalon, midbrain, pons (and cerebellum), and medulla oblongata. Simultaneously, the inside neural canal becomes the ventricles and central canal. The sulcus limitans in the neural canal is the boundary between the sensory nerve and motor nerve, which provides great consistency of neuroanatomy.

Introduction

Life begins the moment a sperm meets an ovum to become a cell.

After the cell goes through countless cell divisions, it eventually becomes a baby. This process is called development.

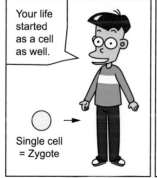

Your life started as a cell as well.

Single cell = Zygote

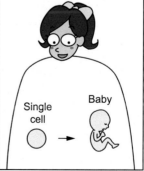

Single cell

Baby

Fig. 5.1

Visually Memorable Neuroanatomy for Beginners. DOI: https://doi.org/10.1016/B978-0-12-819901-5.00005-3

Embryology is the branch of biology that studies the fertilization of sperm and ovum, and the development of embryo and fetus. The main focus is on the "embryonic" stage (8 weeks since fertilization), during which organs including the brain are formed. Therefore, we call the study "embryology."

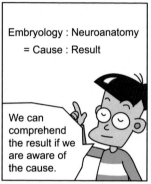

The reason why embryology is covered in this book is that neuroanatomy is the result of embryology.

Embryology : Neuroanatomy
= Cause : Result

We can comprehend the result if we are aware of the cause.

Fig. 5.2

Embryology is very useful in understanding neuroanatomy.

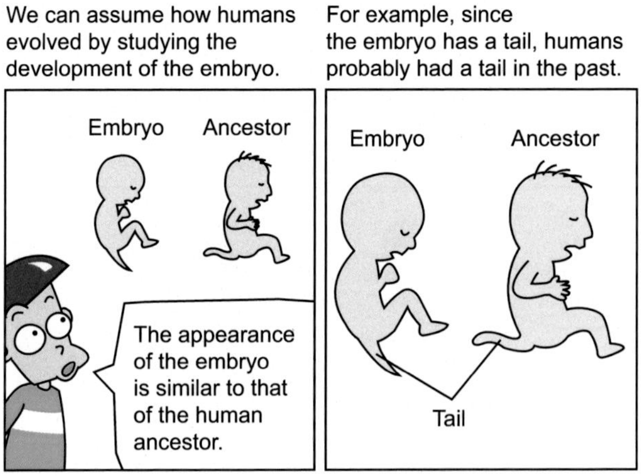

We can assume how humans evolved by studying the development of the embryo.

For example, since the embryo has a tail, humans probably had a tail in the past.

Embryo   Ancestor

Embryo   Ancestor

The appearance of the embryo is similar to that of the human ancestor.

Tail

Fig. 5.3

Since the heart of the embryo consists of one atrium and one ventricle, humans probably evolved from fishes.

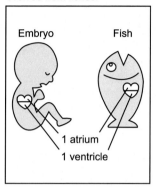

Fig. 5.3 (Continued)

It is believed that embryology and evolution are correlated. The gill-shaped pharyngeal arches (Fig. 3.36) also imply that humans have lived in water and evolved from fishes. A pedantic expression is that ontogeny (developmental process of an organism) recapitulates phylogeny (evolutional process of a species).

Development of the neural tube

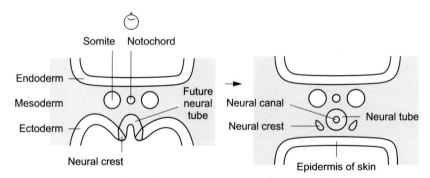

Fig. 5.4 Generation of neural tube.

During the "3rd" week since fertilization, there are "3" layers (endoderm, mesoderm, ectoderm). During the 4th week, the ectoderm generates the neural tube, which will become the central nervous system (brain and spinal cord) (Fig. 1.1).

The ectoderm also develops into the neural crest, which is dorsolateral to the neural tube. The neural crest is destined to become the spinal ganglion and other structures. Depicted here is the spinal ganglion (dorsal root ganglion) that is dorsolateral to the spinal cord (Fig. 3.72).

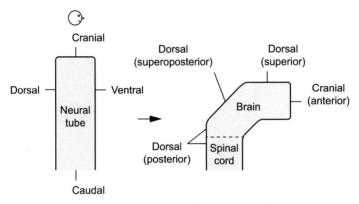

Fig. 5.5 Dorsal, ventral, cranial, and caudal directions before and after head folding.

During the FOurth week, the head FOlding occurs, which causes the neural tube to flex in different angles (Fig. 5.19). The term "dorsal" is constant throughout the neural tube regardless of the flexion angle. Thus, the term "dorsal" is preferred over other confusing terms "superior, superoposterior, or posterior." The same reason is applied to the terms "ventral, cranial, and caudal" in the embryology and neuroanatomy. The term "cranial" is synonymous with the term "rostral."

The amount of neural tube flexion differs between species, but directions of the brain components should be consistently described (for example, both in humans and experimental animals). Therefore, the terms "dorsal, ventral, cranial, and caudal" are preferentially used in comparative anatomy also.

This issue does not apply to the spinal cord, since the spinal cord does not flex during development. That is why the dorsal root (of spinal nerve) (Fig. 3.72) is often called the posterior root.

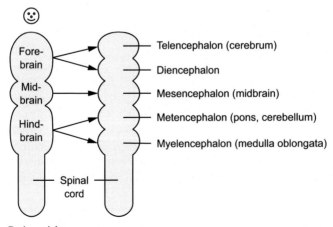

Fig. 5.6 Brain vesicles.

The cranial part (future brain) of the neural tube forms three vesicles: the forebrain, midbrain (mesencephalon), and hindbrain. The "midbrain" is literally the "middle of brain."

Next, the forebrain divides into the telencephalon (cerebrum) and diencephalon, while the hindbrain divides into the metencephalon (pons, cerebellum) and myelencephalon (medulla oblongata) (Fig. 1.11). The remaining caudal part (future spinal cord) of the neural tube does not form a vesicle.

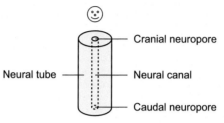

Fig. 5.7 Neural canal.

The initial neural canal in the neural tube (Fig. 5.4) is open both cranially and caudally. These openings, known as the cranial and caudal neuropores, are soon blocked; the cranial neuropore is blocked by the lamina terminalis. The closed neural canal becomes the ventricles and central canal (Fig. 1.11).

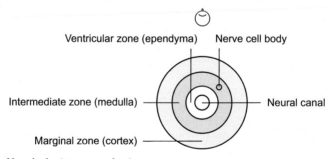

Fig. 5.8 Neural tube (transverse plane).

In the transverse plane, the neural canal and the three zones of the neural tube (ventricular, intermediate, marginal zones) can be identified. The ventricular zone becomes ependyma, lining epithelium of the ventricle (Fig. 1.14). The intermediate and marginal zones develop into the medulla and cortex, respectively. Initially, the medulla contains nerve cell bodies (Fig. 5.9).

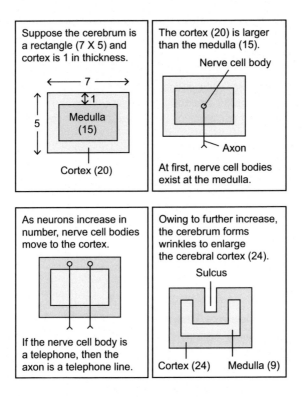

Fig. 5.9

The most cranial part of neural tube is the cerebrum (Fig. 5.6). The nerve cell bodies in the cerebral medulla migrate to the cerebral cortex. Why?

During development, neurons in the cerebrum proliferate enormously. Problem is excessive size of the nerve cell bodies, compared with the axons (Fig. 2.2). To solve this problem, the nerve cell bodies move to the cerebral cortex, which gets even larger in volume after forming the sulci. In the external view of the cerebrum, two-thirds of the cerebral cortex is hidden in the sulci (Figs. 1.31, 1.40).

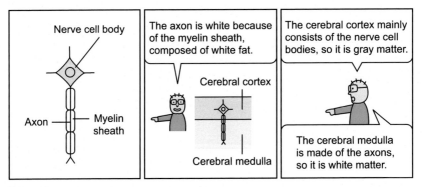

Fig. 5.10

Consequently, the cerebral cortex becomes gray, and the cerebral medulla becomes white (Fig. 1.31). The color difference can be best explained by histology: The axon's myelin sheath is composed of white fat. Think about the white fat in the bacon.

The terms "medulla" and "cortex" to represent location are not changeable (Figs. 1.31, 5.8), but the terms "gray matter" and "white matter" to represent histology are changeable during the development (Fig. 5.9).

Unlike the brain, the spinal cord does not need such a large number of neurons (Fig. 2.23). Therefore, the nerve cell bodies do not migrate, and the gray matter remains inside of the spinal cord (Fig. 1.69). For the same reason, the spinal cord does not get wrinkly (Fig. 5.9).

In summary, the cerebrum's gray matter is external; the spinal cord's gray matter is internal. In the case of the diencephalon, the marginal zone (Fig. 5.8) almost disappears. As a result, the diencephalon such as the thalamus is a mass of nerve cell bodies (nuclei) (Fig. 1.43). The brainstem develops between the cerebrum and the spinal cord (Fig. 4.40); therefore, its nerve cell bodies may or may not move outside. As a result, nuclei and tracts of the brainstem are mixed (Figs. 5.18, 5.21, 5.22, 5.23).

Development of the telencephalon, the diencephalon

The telencephalon grows substantially to become the bilateral cerebral hemispheres. The lamina terminalis is not only the cranial block of the third ventricle, but also the first commissure between the bilateral cerebral hemispheres (Fig. 1.11).

As mentioned, most nerve cell bodies in the cerebral medulla move to the cerebral cortex (Fig. 5.9). However, some nerve cell bodies remain to become the corpus striatum, which is the center of basal nuclei (Fig. 1.37).

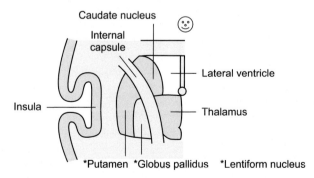

Fig. 5.11 Development of corpus striatum.

The corpus striatum is penetrated by the internal capsule (sensory and motor nerves) to be divided into the lentiform nucleus and caudate nucleus (Figs. 1.39, 1.40). This development is natural because the corpus striatum is located in the cerebral medulla.

The putamen of the lentiform nucleus holds the insula (Figs. 1.27, 1.40) to prevent it from growing outward. So the insula (meaning island) becomes covered by other growing parts of the cerebrum (frontal, parietal, and temporal lobes) to be isolated (Fig. 1.23).

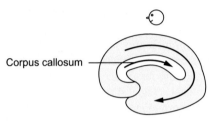

Corpus callosum

Fig. 5.12 Influence by elongation of caudate nucleus.

The caudate nucleus elongates and becomes C-shaped (Fig. 1.39). It determines the C-shaped lateral ventricle (Fig. 1.12), the curved corpus callosum that originates from the lamina terminalis (Fig. 1.44), and the C-shaped cerebrum. Within the cerebrum, the "parietal, occipital, and temporal" lobes are formed in sequence (Fig. 1.23). This developmental order has affected nomenclature, such as the "parietooccipital" sulcus and "occipitotemporal" sulcus (Fig. 1.28).

To sum up, the Striatum (putamen, caudate nucleus) (Figs. 1.37, 1.39) greatly contributes to the Shape of cerebrum. Thus, the striatum can be called the backbone of cerebrum.

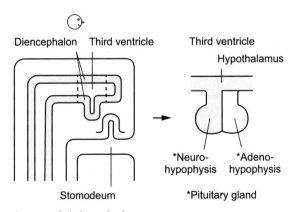

Diencephalon   Third ventricle   Third ventricle

Hypothalamus

*Neuro-   *Adeno-
hypophysis   hypophysis

Stomodeum   *Pituitary gland

Fig. 5.13 Development of pituitary gland.

During development of the diencephalon (Fig. 1.11), its part evaginates ventrally to become the neurohypophysis. That is why the hypothalamus is linked with the neurohypophysis by neuron. Simultaneously, a part of the stomodeum (primitive oral cavity, nasal cavity, and nasopharynx) evaginates dorsally, and develops into the adenohypophysis (Fig. 4.27).

Development of the sulcus limitans

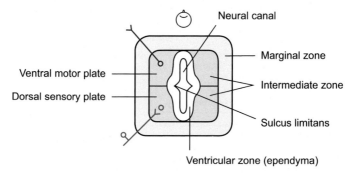

Fig. 5.14   Sulcus limitans of neural tube (transverse plane).

Regarding the three zones of the neural tube (Fig. 5.8), the intermediate zone organizes into the dorsal sensory plate and ventral motor plate. The official terms are the alar plate (for the dorsal sensory plate) and basal plate (for the ventral motor plate); the official terms do not match the orientation of the above transverse plane (Fig. 1.55). Landmark to "limit" the two plates in the neural canal is the sulcus "limitans." Around the sulcus limitans, dorsal side is for sensory nerve, while ventral side is for motor nerve.

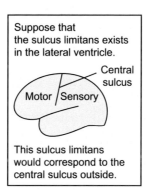

Fig. 5.15

In the cerebrum, the central sulcus is the border between the sensory lobe and motor lobe (Fig. 4.25). If the sulcus limitans is extended to the lateral ventricle, it will correspond to the central sulcus. Actually, the sulcus limitans cannot be seen in the lateral ventricle at all (Figs. 1.12, 1.40).

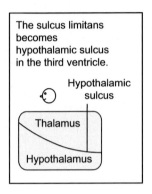

Fig. 5.16

In the diencephalon, the sulcus limitans becomes the hypothalamic sulcus, which extends from the interventricular foramen to the aqueduct of midbrain (Fig. 1.44). This anatomy makes sense because the sulcus limitans runs longitudinally along the lateral wall of the neural canal (Fig. 5.14).

Around the hypothalamic sulcus, the dorsal sensory plate becomes thalamus; the ventral motor plate becomes hypothalamus (Figs. 5.14, 5.17). This is reasonable because the thalamus is confluence of the sensory nerves (Fig. 4.19); the hypothalamus is headquarters of the autonomic nerve (visceral motor nerve) (Fig. 4.28). The exceptions are the ventral lateral and ventral anterior nuclei of thalamus, which are for the motor nerves (Fig. 4.25).

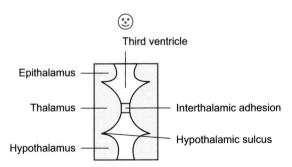

Fig. 5.17 Development of diencephalon (coronal plane).

The diencephalon (Fig. 1.11) develops into the epithalamus (Fig. 1.45), thalamus, and hypothalamus (Figs. 1.44, 4.26, 5.13, 5.16).

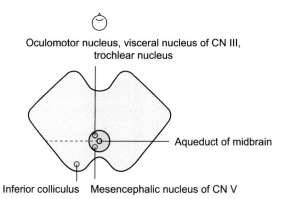

Fig. 5.18 Sulcus limitans (extended to dotted line) (midbrain).

In the midbrain, shape of the original neural canal (Fig. 5.8) remains unchanged; only its name is changed into the aqueduct of midbrain (Fig. 1.11). The aqueduct of midbrain is so small that the sulcus limitans is unrecognizable; but it is imaginable.

The dorsal sensory plate becomes the mesencephalic nucleus of CN V, inferior colliculus (Figs. 3.32, 3.52). The ventral motor plate becomes the oculomotor nucleus, visceral nucleus of CN III, trochlear nucleus (Figs. 3.12, 3.18, 3.23).

This subchapter focuses on the gray matter that is derived from the dorsal sensory plate, ventral motor plate (Fig. 5.14), not on the white matter such as the lemnisci (Figs. 2.11, 2.12, 3.28, 3.52).

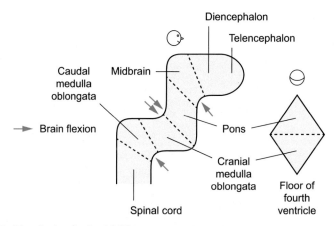

Fig. 5.19 Neural tube after head folding.

During head folding (Fig. 5.5), unexpected reverse flexion happens between the pons and cranial medulla oblongata. The strong reverse flexion (double arrows) induces widening of the neural canal to make the diamond-shaped floor of fourth ventricle (Fig. 1.51). This is similar to a flexed straw of which flexed portion is widened.

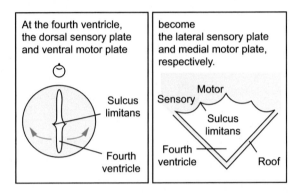

Fig. 5.20

The expanded, and thus thinned, dorsal wall of fourth ventricle becomes its roof, which is composed of the superior and inferior medullary vela (Fig. 1.44). After widening, the sulcus limitans exists between the lateral sensory plate and medial motor plate in the floor of fourth ventricle (Figs. 1.41, 1.54, 1.58).

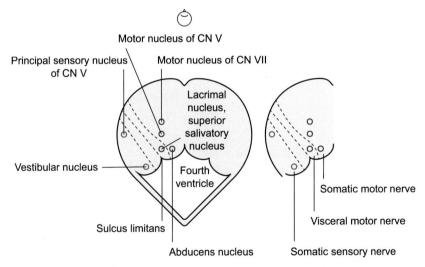

Fig. 5.21   Sulcus limitans (pons).

In the pons, the lateral sensory plate becomes the principal sensory nucleus of CN V, vestibular nucleus (somatic sensory nerve) (Figs. 3.32, 3.45). The medial motor plate becomes the lacrimal nucleus, superior salivatory nucleus (visceral motor nerve) and the motor nucleus of CN V, abducens nucleus, motor nucleus of CN VII (somatic motor nerve) (Figs. 3.25, 3.32, 3.37).

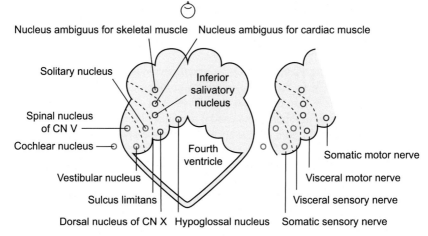

Fig. 5.22 Sulcus limitans (cranial medulla oblongata).

In the cranial medulla oblongata, the lateral sensory plate becomes the spinal nucleus of CN V, vestibular nucleus, cochlear nucleus (somatic sensory nerve) and the solitary nucleus (visceral sensory nerve) (Figs. 3.33, 3.45, 3.52, 3.54, 3.59). The medial motor plate becomes the inferior salivatory nucleus, nucleus ambiguus for cardiac muscle, dorsal nucleus of CN X (visceral motor nerve) and the nucleus ambiguus for skeletal muscle, hypoglossal nucleus (somatic motor nerve) (Figs. 3.55, 3.61, 3.66).

In both the pons and cranial medulla oblongata, the somatic sensory nerve, visceral sensory nerve, visceral motor nerve, and somatic motor nerve are arranged in order. The sulcus limitans is like a mirror between the lateral sensory plate and medial motor plate (Figs. 3.68, 5.21).

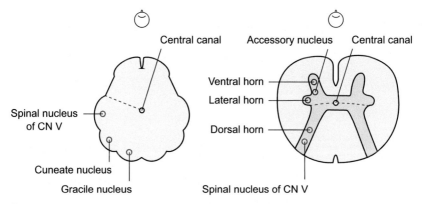

Fig. 5.23 Sulcus limitans (extended to dotted line) (caudal medulla oblongata, left; spinal cord, right).

In the cases of the caudal medulla oblongata and spinal cord, the original neural canal (Fig. 5.8) only changes its name into the central canal (Fig. 1.11), like the aqueduct of midbrain (Fig. 5.18). In the central canal, the sulcus limitans is invisible.

The dorsal sensory plate becomes the gracile nucleus, cuneate nucleus, spinal nucleus of CN V in the caudal medulla oblongata and the dorsal horn, spinal nucleus of CN V in the spinal cord (Figs. 2.11, 2.14, 3.32). The ventral motor plate becomes nothing in the caudal medulla oblongata, but it becomes the ventral horn, lateral horn, accessory nucleus in the spinal cord (Figs. 2.19, 2.28, 3.64).

The sulcus limitans is a keyword to explain many neuroanatomy structures (Fig. 3.68). Other keywords would be the lower and higher levels of nerve activities (Figs. 2.7, 4.4, 4.10, 4.33, 4.34, 4.49) and the general rule of afferent nerves having three neurons (Tables 1, 2, 3).

Tables

Table 1 Three neurons of afferent nerves (first)

Name	Sense	1st neuron		2nd neuron		3rd neuron	
		Start	Ganglion	Start	Decussation	Start	End
Spinothalamic tract	Pain, temperature	Free nerve ending	Spinal ganglion	Dorsal horn	Yes (spinal cord) (→ spinal lemniscus)	Ventral posterolateral nucleus	Postcentral gyrus, paracentral lobule
Medial lemniscus pathway	Touch, proprioception	Encapsulated nerve ending	Spinal ganglion	Gracile, cuneate nuclei	Yes (caudal medulla oblongata) (→ medial lemniscus)	Ventral posterolateral nucleus	Postcentral gyrus, paracentral lobule

Table 2 Three neurons of afferent nerves (second)

Name	Sense	1st neuron			2nd neuron		3rd neuron		
		Start	Ganglion		Start	Decussation	Start	End
CN I (olfactory pathway)	Smell	Olfactory mucosa			Olfactory bulb	No (→ uncus, amygdaloid nucleus, etc.)	(Absent)	(Absent)
CN II (visual pathway)	Vision	Cone, rod cells			Retina	Yes in half (optic chiasm) (→ optic tract)	Lateral geniculate nucleus	Cuneus, lingual gyrus
CN V (trigemino-thalamic tract)	Pain, temperature, touch	Face, etc.	Trigeminal ganglion		Principal sensory, spinal nuclei of CN V	Yes (spinal cord, brainstem) (→ trigeminal lemniscus)	Ventral posteromedial nucleus	Postcentral gyrus
CN VII (taste pathway)	Taste	Taste bud	Geniculate ganglion		Solitary nucleus	No (→ central tegmental tract)	Ventral posteromedial nucleus	Insula, etc.
CN VIII (balance pathway)	Balance sense	Utricle, saccule, semicircular duct	Vestibular ganglion		Vestibular nucleus		Ventral posteromedial nucleus	Cerebral cortex (scattered)
CN VIII (auditory pathway)	Sound	Cochlear duct	Spiral ganglion		Cochlear nucleus	Yes in part (pons) (→ lateral lemniscus) (additional synapse in inferior colliculus)	Medial geniculate nucleus	Transverse temporal gyrus
CN IX (taste pathway)	Taste	Taste bud	Inferior ganglion		Solitary nucleus	No (→ central tegmental tract)	Ventral posteromedial nucleus	Insula, etc.

Table 3 Three neurons of afferent nerves (third)

Name	Function	1st neuron		2nd neuron			3rd neuron	
		Start	Ganglion	Start	Decussation		Start	End
Limbic system (medial limbic circuit)	Memory, emotion	Hippocampus (→ fornix)		Mammillary body	No (→ mammillothalamic tract)		Anterior nucleus	Cingulate, parahippocampal gyri
Basal nuclei (direct pathway)	Appropriate movement	Striatum		Globus pallidus	No		Ventral lateral, ventral anterior nuclei	Precentral gyrus
Ponto-cerebellum (afferent to cerebrum)	Skilled movement	Purkinje cell		Dentate nucleus	Yes (midbrain)		Ventral lateral nucleus	Precentral gyrus
Ascending reticular activating system	Consciousness	Body (whole)	Sensory ganglia (whole)	Reticular formation			Intralaminar nucleus	Cerebral cortex (whole)

Other recommended readings

Blumenfeld H. Neuroanatomy through Clinical Cases. 2nd ed. Sinauer Associates; 2018.

Champney TH. Essential Clinical Neuroanatomy. Wiley-Blackwell; 2015.

Cho ZH. 7.0 Tesla MRI Brain Atlas. In Vivo Atlas with Cryomacrotome Correlation. Springer; 2009.

Chung BS, Chung MS. Homepage to distribute the anatomy learning contents including Visible Korean products, comics, and books. Anat Cell Biol 2018;51:7−13.

Chung BS, Koh KS, Oh CS, Park JS, Lee JH, Chung MS. Effects of reading a free electronic book on regional anatomy with schematics and mnemonics on student learning. J Korean Med Sci 2020;35:e42.

Chung MS, Chung BS. Visually Memorable Regional Anatomy (e-publication at anatomy.co.kr); 2020.

Crossman AR, Neary D. Neuroanatomy: An Illustrated Colour Text. 6th ed. Elsevier; 2019.

Goldberg S. Clinical Neuroanatomy Made Ridiculously Simple. 5th ed. MedMaster Inc.; 2014.

Haines DE. Neuroanatomy Atlas in Clinical Context: Structures, Sections, Systems, and Syndromes. 10th ed. Lippincott Williams and Wilkins; 2018.

Kiernan JA. Barr's the Human Nervous System: An Anatomical Viewpoint. 10th ed. Wolters Kluwer Health; 2013.

Martin JH. Neuroanatomy Text and Atlas. 5th ed. McGraw-Hill Education; 2019.

Snell RS. Clinical Neuroanatomy: Clinical Neuroanatomy for Medical Students. 7th ed. Lippincott Williams and Wilkins; 2009.

Index

Printed in the United States
by Baker & Taylor Publisher Services